"The authors provide an invaluable guidebook to an emerging and at times uncanny technological landscape whose unblinking, opaque, and distributed gaze stares back at us from a growing array of devices that promise to sort, recognize, and evaluate us. *Automating Vision* is a crucial contribution to the new forms of visual literacy we must cultivate if we are to reap the benefits of the burgeoning field of machine vision while evading its pitfalls. It is an elegantly written, theoretically sophisticated book that is destined to become a touchstone work for our times."

Mark Andrejevic, Monash University, Australia. Author of *Automated Media*

"Snapshots are automated, vision becomes machinic, cars sense more than the driver, and seeing is more like data analysis; it's in this field of transformations of media that *Automating Vision* offers an excellent analysis of the social aspects of artificial intelligence. Warmly recommended across the multiple contemporary disciplines that have to make sense of this situation but also to develop a fresh approach to media literacy."

Jussi Parikka, University of Southampton, UK, and FAMU, Prague, Czech Republic

"This timely volume offers a rich discussion of the social impact of smart cameras across a range of domains, ranging from surveillance and facial recognition to drones and self-driving cars. The central term 'camera consciousness' grounds the productive analysis of the social interactions around and with new visual technologies. This book will be a key reference for scholars interested in the social aspects of algorithmic visual technologies."

Jill Walker Rettberg, Author, Professor and Leader of the Digital Culture Research Group at the University of Bergen, Norway

AUTOMATING VISION

Automating Vision explores the rise of seeing machines through four case studies: facial recognition, drone vision, mobile and locative media and driverless cars. Proposing a conceptual lens of camera consciousness, which is drawn from the early visual anthropology of Gregory Bateson and Margaret Mead, *Automating Vision* accounts for the growing power and value of camera technologies and digital image processing.

Behind the smart camera devices examined throughout the book lies a set of increasingly integrated and automated technologies underpinned by artificial intelligence, machine learning and image processing. Seeing machines are now implicated in growing visual data markets and are supported by emerging layers of infrastructure that they coproduce. In this book, Anthony McCosker and Rowan Wilken address the social impacts, the disruptions and reconfigurations to existing digital media ecosystems, to urban environments and to mobility and social relations that result from the increasing automation of vision and explore how it might be possible to ensure a safe and equitable future as we learn to see with and negotiate the interventions of seeing machines.

This book will appeal to students and scholars in media, communication, cultural studies, sociology of media and science and technology studies.

Anthony McCosker is Associate Professor in Media and Communication and Deputy Director of the Social Innovation Research Institute, Swinburne University of Technology, Melbourne, Australia.

Rowan Wilken is Associate Professor in Media and Communication and Principal Research Fellow in the Digital Ethnography Research Centre (DERC), RMIT University, Melbourne, Australia.

AUTOMATING VISION

The Social Impact of the New Camera Consciousness

Anthony McCosker and Rowan Wilken

Routledge
Taylor & Francis Group

NEW YORK AND LONDON

First published 2020
by Routledge
52 Vanderbilt Avenue, New York, NY 10017

and by Routledge
2 Park Square, Milton Park, Abingdon, Oxon, OX14 4RN

Routledge is an imprint of the Taylor & Francis Group, an informa business

Library of Congress Cataloging-in-Publication Data
A catalog record for this book has been requested

ISBN: 978-0-367-35694-1 (hbk)
ISBN: 978-0-367-35677-4 (pbk)
ISBN: 978-0-429-34117-5 (ebk)

Typeset in Bembo
by Apex CoVantage, LLC

CONTENTS

FIGURES

ACKNOWLEDGMENTS

The successful completion of any book is possible only because of the generosity and assistance of many people. In writing this book, we would like to thank Mark Andrejevic, Jussi Parikka and Jill Walker Rettberg for their generous support. Anthony would like to thank Swinburne colleagues who have provided invaluable support and input, in particular Kath Albury, Jane Farmer, Esther Milne, Diana Bossio, Max Schleser, Arezou Soltani-Panah, Roksolana Suchowerska, César Albarrán Torres and Dan Golding. Rowan would like to thank RMIT colleagues who have provided input, assistance and encouragement, including those within the Digital Ethnography Research Centre (DERC) and in the Technology, Communication and Policy Lab, especially Julian Thomas, Hannah Withers and Jenny Kennedy. Others involved further afield through collaboration and events include Jean Burgess, Nic Carah, Terri Senft, Anna Munster, Sarah Tuck, Lila Lee-Morrison, Ysabel Gerrard, Crystal Abidin, Brady Robards, Tama Leaver, Paul Byron, Amelia Johns, Son Vivienne, Nick Davis, Lee Humphreys, Jason Farman and Jordan Frith.

We wish to express our gratitude to Erica Wetter for supporting this book, and to Emma Sherriff and Sarah Pickles at Routledge for shepherding it through to production. We also wish to thank the artists and organizations who have granted us permission to reproduce the images included in this book. Every effort has been made to trace and pay all the copyright holders.

One part of Chapter 5 appeared as "Drone Media: Unruly Systems, Radical Empiricism and Camera Consciousness," *Culture Machine* 16 (2015).

Finally, and closer to home, Anthony would like to thank Leila, Lewis and Edith for their endless support, and Rowan would like to thank Karen, Laz, Max and Sunday for everything.

1

INTERROGATING SEEING MACHINES

If there is a defining image that captures the predicament that smart cameras and machine vision technologies have placed us in, it's this one: early in 2018, reports spread from Chinese government and media sources showing large digital screens at busy intersections pairing face recognition with citizens' identification data, naming and showing the faces of offending jaywalkers.[1] Many of the large cities in China do have problems with traffic congestion and accidents, and so jaywalking may indeed be a serious issue in need of technological redress, but this seems beside the point. One report showed the face of Dong Mingzhu, chairwoman of Shenzhen-listed Gree Electric Appliances, beside her name on a large screen in Ningbo, Jhejiang Province.[2] With this image and others like it, we were introduced to one of the most ambitious – and by most non-Chinese accounts, terrifying – applications of automated governance yet, using smart cameras that combine face recognition with an integrated set of personal identification data to publicize individuals' relatively minor offences as a deterrent.

Pilot programs for China's so called "social credit system" use a range of coordinated technologies, including face recognition, in their camera-centered and increasingly automated governance techniques. The ultimate goal for China is to generate a "trustworthiness index" and, by integrating automated identification, algorithm-assisted policing and social shaming techniques, enter a new phase of social governance and economic order. More generally, in recent years, China has often repeated its aims to become a world leader in machine vision and other artificial intelligence (AI) technologies. While claims of achieving 90% accuracy with social credit–related face recognition applications have been countered as wildly overstated, it is not just the question of accuracy that is at stake in the race to build automated vision systems.

More alarmingly, China has been criticized for using these technologies to target and persecute Uyghur minority populations.[3] We discuss this more in Chapter 3. For now, needless to say, this scenario is a sign that we are in a time of rapid development, experimental application and, perhaps most significantly, wildly divergent understandings of appropriate forms of regulation and control regarding data-driven automated camera technologies.

The networked cameras, real-time image analysis and machine learning systems underpinning the use of face recognition in China's social credit program are part of a group of technologies described as ushering in the "fourth industrial revolution"[4] or industry 4.0 (Figure 1.1). To use the well-known language of the Gartner technology "hype curve," for the majority of industry 4.0 technologies, we are at peak hype and perhaps entering the decline that follows many exemplary failed applications and episodes of negative press. Nonetheless, with massive investment in countries like Germany, China and the United States, the World Economic Forum has characterized industry 4.0 technologies as defining a paradigmatic shift for commerce, industry and human society. Artificial intelligence is the term used to collectively refer to the accelerated developments in data science and machine learning that have improved automated decision-making, supported by integrated cloud computing, alongside advanced robotics, biomechanics and related technologies. As a poor form of shorthand for all of these things and more, AI raises fear and wonder in equal measure.

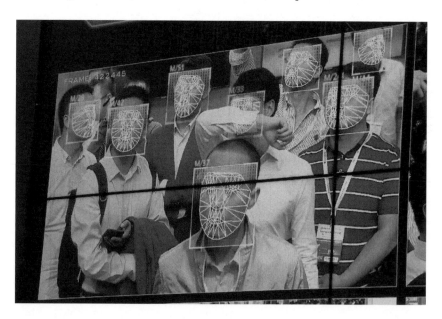

FIGURE 1.1 Chinese AI company Megvii's Face++ face recognition technology presented at a Shenzhen security expo.

Source: Reuters/Bobby Yip.

In this book, we interrogate the social impact of automation and machine intelligence by focusing on the growing applications and implications of smart cameras – or seeing machines. Our core contribution to debates is to point to a particular ambivalence and tension at the heart of automated AI-driven technologies, one that surfaces most dramatically with smart camera and machine vision systems as they make themselves *felt*. With smart cameras, this tension shows up as a new form of heightened camera consciousness. This somewhat anachronistic term, once used to describe the self-conscious effect of cameras when stage performers became cinema actors, holds a new relevance in the age of seeing machines. Camera consciousness designates both the augmenting power and shift in vision and visibility that computer processing contributes to the work that cameras do. It also refers to the multiplying instances of anxiety and vulnerability that follow. Something has shifted in how we make sense of seeing, vision and visibility. In fact, with the proliferation of automated cameras and machine vision systems, we no longer hold the privilege of a self-sustained human-centered concept of *seeing* or *vision* (if there ever was one). There are cameras that see, act and intervene, there are the new relationships of visibility they create, and there are the augmentations and states of anxiety that they affect.

The problem of camera consciousness is one of awareness and attention. Sometimes it's too much, sometimes it's not enough. Of course, it has become commonplace to consider contemporary digitally mediated life to be encompassed by an "attention economy."[5] The duplicitous character of camera consciousness in the age of seeing machines is about the new value in the power to manage that attention, or to automate it and profit from it.

We are dealing, then, with a wider set of debates in the "new dark age" of systemic computation and automation, to use James Bridle's critical language. For Bridle, "computation does not merely augment, frame, and shape culture; by operating beneath our everyday, casual awareness of it, it actually *becomes* culture" (Bridle, 2018: 39).[6] He sees computation as a kind of takeover, an overlay of code and software that remakes the world in its own image. It's worth considering that culture, the social or any interpersonal arrangement has always been mediated. The mixed feelings that computer vision systems bring into play and the contradictory ideas about the magnitude and endpoint of this computational takeover are simply supercharged by the "cloudiness" of networked technologies and the undisclosed intentions of the corporations and governments putting them to use.

Technologically, we are at the point where we can talk about camera conscious machines – machines that gain functionality through their ability to see. Through machine learning vision systems, cameras and the connected seeing machines they inform have attained a kind of awareness that is powerfully felt, and that in turn feels at times oddly powerful and disconcerting or even threatening. The consciousness *of* cameras, our awareness of their disconcerting presence, has always been at the heart of the idea of camera consciousness. And, as we show in Chapter 2, this feeling is often considered a disposition that should be over-come, or else serves a contested point of social etiquette as we learn to be *with* and *before* cameras from an early age. All of the associated technological devel-opments, including machine vision, specialized low-powered image processor chips, broad-spectrum sensors and lenses and cloud computing, combine to boost and exacerbate a reaction that has always been there with cameras.

So, camera consciousness is not just awareness of cameras but also a self-conscious reaction to their power to make visible, to reveal, to capture and hold their target and to fix it for scrutiny, analysis and judgment. Our warning is to not let this sense of camera consciousness recede through the tricks of ubiqui-tous camera systems and barely visible devices that slip into the background. Camera consciousness is something to foster – a skill set of the automated digi-tal age that belongs to the people on the scene as much as the machines. More than this, if we don't attend to the ambivalence that shadows the use of smart cameras – the threats as well as the augmentations they offer – we are sure to embed automated systems with new digital inequalities.[7]

The Ambivalence of Seeing Machines

One interesting element among many in the scenario of China's audacious application of automated camera-powered social credit governance is the role people play in informing the system. There are always reminders that with AI technologies, all is not always as it seems. For one thing, in many instances, "humans still sift through the images to match them to people's identities."[8] A *New York Times* exposé notes the overstated capabilities in achieving China's aim for total algorithmic governance. A photo by Gilles Sabrié accompanying Mozur's article shows dozens of employees of surveillance tech start-up Megvii concentrating their attention on monitors and laptops, playing their part in this "automated" intelligent system as, figuratively speaking, seeing machines.[9] Machine vision systems need big datasets as a starting point that in turn require some form of human input in the form of decision design or "training" – an encoding process that defines the parameters of the automated decisions that follow. The accuracy of the system often depends on the labor of those who establish the layers of classification, the codes that match what the data refers to and the algorithms that underpin its mechanisms of sensemaking. The banks of monitors and human labor powering Megvii are a reminder at the outset that machines can be designed to see, decide and act but not without human guidance and multiple divergent interests. Such revelations of human input and oversight, ironically, introduce both reassurance and doubt.

At every point in the development and application of machine vision, a deep ambivalence defines the forking possibilities of its use. This is cause for inter-est and concern. Every camera augments human seeing, and that's the attrac-tion. While surveillance and automated governance are important to the big brother-like scenario signaled in China's application of face recognition, there are many other facets to the way machine vision technologies are beginning to change society. Throughout this book, we look at how machine vision augments and reshapes the social through both spectacular and mundane applications, all of which affect the dynamics of vision and visibility. Whether we're talking about the cameras that publicly shame jaywalkers at a busy intersection, or guide driverless cars, or a drone seeing and avoiding obstacles while making its own way "back to base," *ambivalence* shapes the underlying tensions between eager investment and generalized anxiety. By centering our discussion of machine vision technologies and their applications on the concept of the new camera consciousness, we aim to probe in every case the directions this ambivalence can lead us.

Technological ambivalence is often stated up front in accounts of new things as diverse as drones or machine learning systems. The idea is that the technology itself is neutral, that it's up to "us" – but who exactly? – as to how it is used or applied and to ensure that it is applied fairly, safely and with human and social interests in mind. In the traditions of science and technology studies, this ambivalence has been understood as central to the ebbs and flows of many types of technological development. In popular debates about contentious technologies like face recognition, however, it can underpin and build social movements around fear and doubt or be used to challenge public concerns as simply hindering "progress" or jeopardizing safety and security. These debates are rarely helpful by themselves. Rather than asking "are these technologies inherently good or bad?" it is better to know whether or how they can be used to improve lives and understand the specific circumstances in which they do not. The social science of machine vision is in its infancy and often serves as an aside, a critical point of view, or in the service of overcoming the "acceptance problem." Media and communication technologies have always shown a transformative capacity, but their application for social good never comes without hard fought contest.

We are increasingly losing ownership of the act of seeing and sensemaking. There is, as Mark Andrejevic puts it, a cascading logic to automation, where "automated data collection leads to automated data processing, which, in turn, leads to automated response."[10] The cascade is not necessarily reductive. Certain elements of these processes were conceded long ago – through the mechanisms of artistic painting and perspective, personal photographic cameras, cinema and video. The difference is that cameras have traditionally produced images for consumption or for human view. Machine vision renders the world not for human eyes but to power automated monitoring, analysis and decision-making. The biggest loss in this step is transparency and a sense of reciprocity in the relationships of vision and visibility, of what it means to see and be seen. Are we consigned to living with "seeing machines" that are "black boxes" of algorithmic decision-making, to use Frank Pasquale's terminology?

Taina Bucher makes the point that despite the prevailing sense that algorithms and machine learning systems are opaque at best, or even by definition operate outside of human reasoning, we can and should appraise them through their effects and impacts. The goal is to maintain scrutiny and pressure on those who design and implement these systems but also to redirect their goals. Visual data has a newfound value. There is an urgent need to set our focus on the machines, infrastructures, devices, programs and systems that capture and realize that value and power.

One of the most useful counterpoints for the ambivalence of complex and contentious technology is transparency. Transparency could even be considered a core tension in the evolving social arrangements brought about by the technologies of the new camera consciousness. Machine learning techniques produce results through processes rarely known even to their creators. Just the idea of a camera that is able to both see *and* judge intensifies the breakdown of transparency and reciprocity that cameras have always to some extent triggered. The twentieth century established portable cameras as powerful witness in times of war, natural disaster or political scandal, providing intimate windows to personal and social lives. Automation and machine vision are a logical and qualitative step away from these recording and representational functions. It is one thing to lose sight of cameras as they saturate urban and private spaces, but there are significant consequences if we fail to understand and regulate their new capacity to judge, discern, distinguish, decide and intervene. As automated vision shifts the locus of control of seeing to distributed camera viewpoints, there is work to do in mapping the reconfiguration of a social order ostensibly built on relationships of visibility and awareness. Transparency is the battleground. But so is expertise.

It could be argued that certain types of specialized seeing have long been ceded to a more select group in our societies. Doctors, physicians, psychologists all make judgments and diagnoses on the basis of specialized observations. Sometimes these modes of specialized seeing are machine assisted, sometimes they are not. A key instance of this, not covered directly in this book, is the impact of machine vision on medical diagnosis and decision-making for therapy or treatment. These applications open up a space for thinking about the potential benefits of machine vision. In fact, medical imaging has been quite an unsung hero over the past century, enabling treatment by making anomalies precisely visible and measurable and hence actionable. Trained medical experts use imaging to diagnose and plot treatment options, but often with images that are incredibly difficult to read. Adding machine vision to the mix, to speed up and even improve that detection–diagnosis work feels right.

Australia and New Zealand have the highest rates of melanoma, the aggressive malignant skin cancer, worldwide. Since the 1990s, machine vision and AI techniques have been used to improve its early detection or monitor changes in affected skin. The problem in this context is the latent space, the hidden space between finding "normal anomalies" and the kind of anomalies that require action. If, as Thomas Schlegal and colleagues claim, "relevant anomalies can be identified by unsupervised learning on large-scale imaging data," then why not make use of AI techniques for diagnostic machine image processing?[11] Dermatologists are trained to understand the fine changes at the edges of moles or skin lesions. They have visual expertise that determines which are the changes that require a biopsy for further cell-level testing or surgery for removal. There is little question of the benefit of improving these processes with machine vision techniques. People in places without access to adequately trained specialists may have a chance of early detection and lifesaving surgery. Although the techniques are not yet fully developed, and comparative studies that pit dermatologists against machine vision systems have been mixed, the signs are good for augmented melanoma diagnosis.[12]

For decades, machine vision techniques have been trained for inspection and measurement in a wide range of contexts. In construction and manufacturing, machine vision is increasingly applied using the same basic techniques as those targeting anomalies in medical images. The systems are technically not all that far from those of automated surveillance systems.[13] In each of these contexts, however, the ambivalence that gives pause for thought is built on the new exchange of trust they *impose* upon the people most directly affected. Those who are crossing a bridge that has been machine-inspected to detect hidden faults, or place their faith in machine vision medical diagnosis designed to find early-stage cancer, are in a similar position to those subject to policing or state governance through image monitoring and face recognition. Each is placed in a new kind of relationship of trust in the workings of a complex system.

How people and machines adapt to the conditions of the new camera consciousness involves a redistribution of seeing among classes of experts, technical systems, hardware, cloud computing and wireless infrastructures, and among new regulatory frameworks and norms. Without wishing to overstate our claims too soon, these add up to substantial change to the fabric of the social.

New Literacies

It's not enough just to know how things work. As James Bridle puts it, "what is required is not understanding, but literacy."[14] Bridle's account of what this literacy would look like for AI technologies is squarely focused on critique. True literacy, he argues, includes comprehension of context and consequences, the interrelationship between systems and their limitations, fluency in the language of the system but also its metalanguage or how it talks about itself and interacts with other systems. Literacy, he says, "is sensitive to the limitations and the potential uses and abuses of that metalanguage."[15] Overall, for Bridle, AI literacy involves the ability to perform and respond to critique. We think that the kind of literacy needed to understand, negotiate, shape and live with machine vision systems should be more than this. Throughout the chapters of this book, we consider the involvement of camera operators, mobile device and app users, drivers, pilots, developers and others in the coproduction or negotiation of automated vision systems. The final chapter turns more directly to the question of the new visual literacies that accompany these systems or are required if we are to adapt socially and personally to the conditions of the new camera consciousness.

As a starting point, all of the relatively mundane contests over language play a vital role in shaping the way new technologies affect human lives. Some of the key forums for language building in relation to automated vision systems take place in government reports, patent applications, technology news, investigative journalism, developer blogs, code repositories like GitHub and similar sites. These are our weather vanes in a time of "interpretive flexibility" to use Pinch and Bijker's framework for understanding the domestication and social shaping of technology.[16] Pinch and Bijker were emphasizing the early divergent and contested meanings that follow the introduction of new technologies but also the competing and often malleable purposes to which new technologies can be put. At this point in time, we are in great need of better conceptual language and attention to the many different technologies, social contexts and applications associated with automated visual systems.

The complexity of the components of what is broadly thought of as AI can itself lead to problems of ignorance or even deliberate mystification. Most primers on AI will start with the caveat that, when companies say they are using AI to do this or that, they're often not. Recommendation systems, autoreply mechanisms and classifiers aren't always achieving what might be considered a level of artificial thinking or general awareness. In fact, the goal of "artificial general intelligence" is a long way off. It's the primary mission of the multibillion-dollar research investment of the OpenAI Institute, for instance. OpenAI defines AGI (artificial general intelligence) as "highly autonomous systems that outperform humans at most economically valuable work."[17] While significant gains continue to be made toward this goal, there is still a long way to go. What we have instead are a collection of more or less accessible, more or less open-access and open-source or, on the other hand, obscure, "blackboxed," individual computational systems. These are specialist algorithms and data processing techniques that work alongside dedicated hardware (chipsets, special cameras and sensors) to enable augmented and sometimes automated vision and decision-making tasks. These technologies work to sensibly sort, classify, find patterns and help to make decisions in ways analogous to what humans do all the time. They use, however, the digital data fed through networked sensing systems and online collections.

A wealth of books, online courses, software packages and manuals, university courses and the like provide introductions to machine learning and machine vision.[18] Our brief précis of some of the core elements merely aims to set the scene. Mostly what is referred to as AI involves some form of machine learning system. Machine learning is a set of computer tools and techniques that allow a program to automatically improve its performance on a task, usually by drawing lessons from a large range of similar data. Rather than giving the program a complete set of instructions, as you would with a traditional algorithm, a machine learning algorithm works out its own constraints or pathways for solving a problem with varying degrees of human "supervision" or training. Some machine learning systems are supported to achieve accurate results through human-labeled or carefully tailored input data and prespecified "best case scenario" output data. These are supervised systems. With all the data available, and a clear method and end-goal, the systems are designed to find their own solutions. An unsupervised machine learning system doesn't use a desired best-case scenario or solution. It infers patterns or objects from a dataset without reference to known or labeled outcomes. It might be a system for classifying things according to how the machine thinks they best relate. Or it could have a stack of images of meals, for example, and a set of ingredients and sets out through a number of layered steps to find the meal that is most likely to correspond.

Where machine learning techniques really took off is with the development of deep learning using artificial neural networks. Neural networks are algorithms that use nonlinear processes to sort and learn from very large datasets. As the name suggests, they are brain-like systems, but it's more accurate to say that neural networks in computation are inspired by some of the things known about how information processing happens in the brain. What deep learning achieves is a layering of the experiential steps that a computer takes to resolve problems. The nonlinear processing and large datasets mean that they can extend beyond the features fed in as inputs or guides, to learn underlying structures, patterns and abstractions that might otherwise be missed even through supervised machine learning techniques.

Automated machine learning, or AutoML, is perhaps where the trouble starts. AutoML is nominally about making machine learning accessible, democratizing its capabilities for those unable to negotiate the complex work that specialists undertake to establish a functioning system. It's about creating "point-and-click" interfaces to "plug into," to develop applications in the world, so that all can "reap the benefits" from machine learning. This means automating a number of complex, iterative steps that go into making a machine learning system fit for purpose. There are open-source packages and projects for making AutoML accessible, such as Auto-Keras. However, AutoML could be seen as the business model that enables AI development at scale, and in doing so both draws on and grows the value invested in data and algorithm work for the big technology companies, including Google (Cloud AutoML), Amazon (Amazon Web Services) and Microsoft (Azure Machine Learning). Although these developments may be premised on openness and improving the accessibility of the technology, they contribute to the bundling or even hoarding of assets – algorithms and training data – as the primary value of buy-in services. These moves create a reliance on those well-resourced infrastructures, while reducing transparency and accountability or capacity to regulate because of the size and power of the big technology corporations.

These developments increase the urgency for ongoing scrutiny. There are some signs that critical accounts of machine vision systems – especially for face recognition – have gained traction. Calls for bans on the use of face recognition in policing, on the grounds of unreasonable intrusions on the rights of citizens, inherent biases and unacceptable error rates, have forced political caution in some contexts at least. Kelly Gates' work on the rise of face recognition as part of a "biometric future" increasingly tied to AI set out both a history of its technical development and clear warnings about a new security-conscious digital citizenship.[19] Likewise, the wider field of critical data and information studies has doggedly pursued the inequalities and dangers that plague algorithmic and automated systems.[20] Similarly, a growing field of scholarship is beginning to highlight the often-unrecognized influence of the social values that become embedded in the design or training of AI systems. "Algorithmic auditing" is a burgeoning set of research and advocacy interventions that seek to provide external validation or critical accountability for the effects, accuracy and consequences of algorithms and autonomous AI systems like face recognition platforms. Algorithmic auditing can involve technical studies of accuracy and bias in automated systems,[21] but it can also involve critical input at many levels of the construction and application of those systems, including through more accessible ethnographic methods[22] or by placing the burden of transparency on AI service providers themselves.[23]

Maintaining scrutiny, demanding transparency and consumer protection is tough. Members of the nonprofit organization Fight for the Future set up an interactive map of where across the United States face recognition technologies are being used and resisted.[24] Their map identifies places where Amazon have formed partnerships with police departments for those departments to encourage the use of Amazon's aggressively marketed home security system Ring, in exchange for access to the face data produced. Advocacy of this kind can cut through and aid other forms of political and organizational intervention.

In their Excavating AI project, Kate Crawford and Trevor Paglen attempt to pull back on the reigns of the automation of machine vision systems to ask the question, "what sorts of politics are at work when pictures are paired with labels, and what are the implications when they are used to train technical systems?"[25] When AutoML services are streamlining the construction, dataset management and framework components of ML systems, there is more need than ever to unstitch the packages that underpin the judgments, choices and hence consequences of AI. Crawford and Paglen describe their project as an "archeology of datasets,"[26] an attempt to reveal the "underlying logic of how images are used to train AI systems to 'see' the world." All interventions that test or explicate the workings of automated systems, or establish new angles for insisting on transparency and accountability, play an important part in involving people in the decisions that lead to the specific impact these technologies will have on our lives.

The Social Science of AI

In the absence of sustained examples and case studies, we are left to imagine the benefits and possible long-term consequences of technological systems that replicate human visual awareness but work outside of human view. Ironically, machine vision systems are being put into service with little oversight. The kind of seeing machines that automate governance in the Chinese social credit system might be sufficiently dramatic to inspire popular concern, but the applications are often more mundane. The future of machine vision includes security, surveillance and governance systems, but it also involves the more incremental and invisible reorganization of cities, transport, construction, manufacturing and maintenance. It will affect the hospitality industry, add new assistive technologies in homes or hospitals and push for new modes of economic productivity, labor efficiency and improved health and well-being. Nonetheless, in many cases, machine vision systems have not achieved a "social license to operate." They may realize degrees of legitimacy, but often waver on both credibility – whether or not they work – and especially trust. This is part of the problem that social science and humanities research can help address. While we might recoil from the use of machine vision to publicly shame jaywalkers, for example, its ability to improve the detection and monitoring of disease progression should give pause for thought.

Machine vision needs research approaches that can prize open the changing nature of digital images and the extra-human status of visibility and scrutinize mobility as a way of contributing to AI's appropriate application for the social good or even for ethical commercial profit. It's always tempting to jump to the critical extremes. What's needed is an approach to research and advocacy that broadens inclusion and involvement in assessing the value of AI and looks for pathways toward *inclusive AI*. Everyone should be able to contribute their experiences to the kinds of critical and responsive thinking required both to help develop inclusive, safe, ethically grounded AI vision systems and as a way of participating actively in the social effects and benefits they might offer. An important step involves overcoming the human-centric study of technology that misses the expanding agency of the machines and media that surrounds us, while maintaining a sharp eye on the human, corporate and state *interests* that are also shaping those technologies along well-worn profit lines. Equally important is a scrutiny of the commercial interests at play in AI and machine vision systems, whether in the way they define the criteria for automated credit checks,[27] extract and compile the sample face image data to train machine learning systems and other algorithmic sorting machines[28] or intervene in social platforms through moderation and automated recommendations.[29]

Our book explores seeing machines through four main case studies: face recognition, drone vision, mobile and locative media and autonomous vehicles. Behind the smart camera devices that we explore lies a set of increasingly integrated and automated technologies underpinned by AI, pushing the boundaries of machine learning and image processing. Seeing machines are implicated in growing visual data markets, supported by the emerging layers of infrastructure that they coproduce as images are poured into machine learning databases to improve their accuracy. Our focus lies in understanding the social impact of these new systems of seeing and machinic visuality, the disruptions and reconfigurations to existing digital media ecosystems, to urban environments and to mobility and social relations. Chapter 2 details the concept of camera consciousness and positions it as a means of accounting for the social changes entailed by the increasing autonomy and intelligence of camera technologies and machine vision. We begin with the visual anthropology of Gregory Bateson and Margaret Mead, who pioneered the use of cameras to discover the otherwise unseeable cultural and personal traits of their Balinese subjects. The conceptual framework for camera consciousness as a tool for interrogating machine vision borrows from the theoretical work of William James, Gilles Deleuze, Jane Bennett and other contemporary theorists – each of whom helps us to rethink the materiality of the image and the pragmatics of the relationships that they affect.

In Chapter 3, we address the value of the face as data in the application and social context of face recognition technology. Expanding on the example of China's camera-driven social credit governance system, this chapter offers an expanded account of the camera's surveillant functionality. Built on technologies of recognition and matching, identification and sentiment analysis, the long coveted visual data of the face offers a wide range of possible applications. Just as social media selfies become relatively ubiquitous and domesticated as everyday communicative practice, the value in the visual data of faces is unfurling. In fact, face recognition technology makes most sense, and has the greatest impact, in a time of social media and in relation to the depth of digital traces associated with the media accounting[30] of personal Internet and mobile phone use. Selfies, for instance, far from being the perfect negative trope of the narcissistic mobile and social media age, present a powerful base layer of personal data to be mined, combined and operationalized for any number of intelligent and automated services based on computing individuals and populations.

Chapter 4 considers the automating and augmenting of mobile vision capture. It opens with a historical account of key steps and stages in the development of camera automation and of the incorporation of cameras into mobile phones. Then, in the second part of the chapter, the emphasis shifts by looking beyond the predominant focus on camera phone practices in studying the application of current mobile media to explore the autonomous production of information and metadata, including geo-locative metadata, that accompany and are left behind by camera phone and smartphone use. We shift the emphasis from acts of photography to the relative autonomy of mobile imaging and the increasing value of geo-located visual data.

Augmented reality has been a slow burn application for smartphones, finding a range of uses for the mobile camera as interface or overlay. In Chapter 4, we explore the transition toward a more embedded, everyday activation of augmented reality. Mobile imaging and visual processing technologies play an interesting role in reshaping everyday visibilities. We can see this in developments in internal mapping as it is imagined and tested through Google's Project Tango mobile devices. This project is emblematic of the trajectory of personal mobile media involving sophisticated sensing and visual processing power. As technologies of the new camera consciousness, mobiles have for some time now exemplified the expanded capacity to *see* through distributed points of view; added to this capacity are intelligent cameras, AI image processing and associated sensors that are networked and coupled with machine vision and cloud computing. In this sense, mobiles provoke questions about *how* individual local environments become visible and socially available. They also extend possibilities of seeing and acting digitally beyond human senses.

Drone vision, the focus of Chapter 5, presents a literal liftoff point for considering the motile camera – the camera in self-sustained and computer-controlled aerial movement. Chapter 5 asks: how does the aerial drone affect and depend on new forms of distributed, mobile, wireless visuality? While a lot of attention has been paid to the trajectory of drone technology as it has arisen out of military contexts, we focus on the essential role played by drones' camera technology, visual controls and wireless relations. To understand the implications of drones in specific functional contexts – urban environments, manufacture and maintenance, agriculture, mining and forestry, for instance – we look at how they reconfigure personal, public and environmental visibility, or what it means to see and be seen. Among other things, drone vision pushes the boundary of, and tests new experiences of, what is considered "public." The chapter examines the semiautonomous vision technology that allows drones to act in the world. It explores the "creepy agency" of drone cameras as a factor of the new visual knowledge they are able to produce and probes the meaning of "autonomy" for these machines that act *like* insects or in their unpredictable movements and their ability to swarm or to take the position of the "fly on the wall" but also in their occasional waywardness. We argue that drone vision presents us with a key case of altered sociality triggered by a highly contested camera consciousness.

What can an autonomous vehicle see? Chapter 6 examines a little understood aspect of autonomous vehicle technology: vision capture and image processing. While most attention relating to driverless cars has revolved around legal and safety concerns associated with removing a driver's control, fundamental to their long-term success is the development of integrated visual capture and processing – that is, the processes that are involved in mapping, sensing and real-time dynamic visual data analysis within dynamic urban environments. To become properly integrated into urban, suburban or country transport systems, autonomous vehicles have to adapt dynamically in response to the way those spaces, and all their component objects, obstacles and subjects, can be seen or rendered visible, mapped and understood computationally as real-time traversable data. In examining the vision capture and processing used by autonomous vehicles, we ask how a driverless car *learns* to see, and we explore the human and social challenges that visual processing technologies place in the path of autonomous vehicle developers. The chapter also considers the concept of technological affordances, as set out by James Gibson, in relation to a wholly complex ecosystem. Driverless cars stretch the sense in which a machine can be defined by its affordances for seeing and acting in its environment.

The final chapter points toward what a digital, visual, media data literacy might look like in the age of automation and AI. We adapt the concept of "camera consciousness" as a tool to canvas the wide range of technologies and contexts that are the battlegrounds for AI and to find a space to build new digital, visual and data literacies in the age of seeing machines. Digital literacy has been used as a broad term for addressing both the technical skills and the "cognitive and socio-emotional aspects of working in a digital environment," as Eshet-Alkalai put it in 2004.[31] If we take the case of deepfake videos – videos that use deep learning techniques to artificially construct videos of people acting and speaking in ways that look like the real thing – the time has come to extend our understanding of literacy as read-write skills to the capabilities of agentic seeing machines. A key avenue for shaping the ethical and beneficial development of machine vision technologies lies at the intersection of politics, education and responsible design. To this end, machine vision needs to be accountable and responsive to its own semiotic processes, its meaning-making and sensemaking actions and interventions in the world.

Notes

1. Meng Jing, "From Travel and Retail to Banking, China's Facial-recognition Systems Are Becoming Part of Daily Life," *South China Morning Post*, February 8, 2018, www.scmp.com/tech/social-gadgets/article/2132465/travel-and-retail-banking-chinas-facial-recognition-systems-are.
2. Li Tao, "Face Recognition Snares China's Air Con Queen Dong Mingzhu for Jaywalking, But It's Not What It Seems," *South China Morning Post*, November 23, 2018, www.scmp.com/tech/innovation/article/2174564/facial-recognition-catches-chinas-air-con-queen-dong-mingzhu.

3. Paul Mozur, "One Month, 500,000 Face Scans: How China Is Using A.I. to Profile a Minority," *New York Times*, April 14, 2019, www.nytimes.com/2019/04/14/technology/china-surveillance-artificial-intelligence-racial-profiling.html.
4. Klaus Schwab, "The Fourth Industrial Revolution: What It Means, How to Respond," *World Economic Forum*, January 14, 2016, www.weforum.org/agenda/2016/01/the-fourth-industrial-revolution-what-it-means-and-how-to-respond/.
5. Tiziana Terranova, "Attention, Economy and the Brain," *Culture Machine* 13 (2012), https://culturemachine.net/wp-content/uploads/2019/01/465-973-1-PB.pdf.
6. James Bridle, *New Dark Age: Technology and the End of the Future* (New York: Verso Books, 2018), 39.
7. Virginia Eubanks, "Trapped in the Digital Divide: The Distributive Paradigm in Community Informatics," *The Journal of Community Informatics* 3, no. 2 (2007), http://blog.ci-journal.net/index.php/ciej/article/view/293; Virginia Eubanks, *Automating Inequality: How High-tech Tools Profile, Police, and Punish the Poor* (New York: St. Martin's Press, 2018); Andrew Feenberg, "The Ambivalence of Technology," *Sociological Perspectives* 33, no. 1 (1990): 35–50.
8. Paul Mozur, "Inside China's Dystopian Dreams: A.I., Shame and Lots of Cameras," *New York Times*, October 7, 2018, www.nytimes.com/2018/07/08/business/china-surveillance-technology.html.
9. Ibid.
10. Mark Andrejevic, *Automated Media* (New York: Routledge, 2019), 9.
11. Thomas Schlegl, et al., "Unsupervised Anomaly Detection with Generative Adversarial Networks to Guide Marker Discovery," in *Information Processing in Medical Imaging. IPMI 2017. Lecture Notes in Computer Science, vol. 10265*, Marc Niethammer, et al. (Eds.) (Cham, Switzerland: Springer, 2017), 146–157.
12. Michael A. Marchetti, et al., "Results of the 2016 International Skin Imaging Collaboration International Symposium on Biomedical Imaging Challenge: Comparison of the Accuracy of Computer Algorithms to Dermatologists for the Diagnosis of Melanoma from Dermoscopic Images," *Journal of the American Academy of Dermatology* 78, no. 2 (2018): 270–277; Hiam Alquran, et al., "The Melanoma Skin Cancer Detection and Classification Using Support Vector Machine," in *Proceedings of the 2017 IEEE Jordan Conference on Applied Electrical Engineering and Computing Technologies (AEECT)* (New York: IEEE, 2017), 1–5, https://doi.org/10.1109/AEECT.2017.8257738); Arve Kjoelen, et al., "Performance of AI Methods in Detecting Melanoma," *IEEE Engineering in Medicine and Biology Magazine* 14, no. 4 (1995): 411–416; Samy Bakheet, "An SVM Framework for Malignant Melanoma Detection Based on Optimized Hog Features," *Computation* 5, no. 4 (2017): 1–13.
13. See, for example: Je-Keun Oh, et al., "Bridge Inspection Robot System with Machine Vision," *Automation in Construction* 18, no. 7 (2009): 929–941.
14. Bridle, *New Dark Age*, 3.
15. Ibid.
16. Trevor J. Pinch and Wiebe E. Bijker, "The Social Construction of Facts and Artefacts: Or How the Sociology of Science and the Sociology of Technology Might Benefit Each Other," *Social Studies of Science* 14, no. 3 (1984): 399–441.
17. "About OpenAI," *openai.com*, https://openai.com/about/.
18. See, for example: Towards Data Science: Sharing Concepts, Ideas, and Codes, *Medium*, https://towardsdatascience.com/; Stanford University's Coursera Machine Learning Course, www.coursera.org/learn/machine-learning; GitHub's collection, Get Started with Machine Learning, https://github.com/collections/machine-learning; Jan Erik Solem, *Programming Computer Vision with Python: Tools and Algorithms for Analyzing Images* (New York: O'Reilly Media, Inc., 2012).
19. Kelly A. Gates, *Our Biometric Future: Face Recognition Technology and the Culture of Surveillance* (New York: NYU Press, 2011), 126.

20. Viktor Mayer-Schönberger, *Delete: The Virtue of Forgetting in the Digital Age* (Princeton, NJ: Princeton University Press, 2011); Andrejevic, *Automated Media*; John Cheney-Lippold, *We Are Data: Algorithms and the Making of Our Digital Selves* (New York: NYU Press, 2018).
21. Christian Sandvig, Kevin Hamilton, Karrie Karahalios, and Cedric Langbort, "An Algorithm Audit," in *Data and Discrimination: Collected Essays*, Seeta Peña Gangadharan, Virginia Eubanks, and Solon Barocas (Eds.) (New York: New America, Open Technology Institute, 2014), 6–10; Inioluwa Deborah Raji and Joy Buolamwini, "Actionable Auditing: Investigating the Impact of Publicly Naming Biased Performance Results of Commercial AI Products," in *Proceedings of the 2019 AAAI/ACM Conference on AI, Ethics, and Society (AIES-19)* (New York: ACM, 2019), https://dam-prod.media.mit.edu/x/2019/01/24/AIES-19_paper_223.pdf.
22. Nick Seaver, "Algorithms as Culture: Some Tactics for the Ethnography of Algorithmic Systems," *Big Data & Society* 4, no. 2 (2017), https://doi.org/10.1177/2053951717738104.
23. Brent Mittelstadt, "Automation, Algorithms, and Politics| Auditing for Transparency in Content Personalization Systems," *International Journal of Communication* 10 (2016): 4991–5002, https://ijoc.org/index.php/ijoc/article/view/6267/1808.
24. Fight for the Future, "Ban Face Recognition," www.banfacialrecognition.com/map/; Jillian D'Onfro, "This Map Shows Which Cities Are Using Face Recognition Technology – And Which Have Banned It," *Forbes*, July 18, 2019, www.forbes.com/sites/jilliandonfro/2019/07/18/map-of-facial-recognition-use-resistance-fight-for-the-future/#40de1d587e61.
25. Kate Crawford and Trevor Paglen, "Excavating AI: The Politics of Images in Machine Learning Training Sets," *excavating.ai*, September 19, 2019, https://excavating.ai.
26. Ibid.
27. Eubanks, *Automating Inequality*; Safiya Umoja Noble, *Algorithms of Oppression: How Search Engines Reinforce Racism* (New York: New York University Press, 2018).
28. Kelly A. Gates, *Our Biometric Future: Face Recognition Technology and the Culture of Surveillance* (New York: New York University Press, 2011); Mark Andrejevic and Kelly Gates, "Big Data Surveillance: Introduction," *Surveillance & Society* 12, no. 2 (2014): 185–196.
29. Tarleton Gillespie, *Custodians of the Internet: Platforms, Content Moderation, and the Hidden Decisions That Shape Social Media* (New Haven, CT: Yale University Press, 2018).
30. Lee Humphreys, *The Qualified Self: Social Media and the Accounting of Everyday Life* (Cambridge, MA: MIT Press, 2018).
31. Yoram Eshet-Alkalai, "Digital Literacy: A Conceptual Framework for Survival Skills in the Digital Era," *Journal of Educational Multimedia and Hypermedia* 13, no. 1 (2004): 93–106.

2

CAMERA CONSCIOUSNESS

In a book published in 1942, Gregory Bateson and Margaret Mead developed what they saw as "experimental methods" to reveal previously unrecorded characteristics of the lives of the people of Bali.[1] Their work became a classic of visual anthropology for its methodical use of photography and film to let the image bridge the gap between science and art in communicating the lives of our distant others. As Ira Jacknis explains, this was more than just using images or film for illustration.[2] In their search for an objective record, Bateson and Mead formulated a shift from a verbal, language-based study of culture to using visual data as the basic unit of analysis. Explaining the method, Bateson and Mead note the challenges and potential of using cameras and photographs in "stating the intangible relationships among different types of culturally standardizable behaviour."[3] They explain how the camera can become a tool for solving "problems of description and analysis"[4] and, in doing so, represent an early exercise in a broader trajectory of the anthropology and science of visual analysis based in the extraction of information and data to produce new knowledge.

Throughout the second half of the twentieth century, the analytical use of cameras would develop beyond the subdiscipline of visual anthropology, in satellite mapping and remote sensing, camera-based surveillance, visual profiling and more contemporary techniques of face recognition, along with medical bioimaging, street view mapping, and so on. Through digital formats and networked distribution, cameras have become ubiquitous and increasingly autonomous, distanced from human photographers but also multiplied by mobile device. Most significantly, they have become intelligent through advances in real-time cloud-processed machine vision underpinned by machine learning. The integration of sophisticated forms of camera sensing and digital image processing has jumpstarted new ways of knowing and acting on the world.

As a way of understanding how camera technologies extract knowledge and other forms of value from visual data, particularly their vital role in emerging technologies like drones and autonomous vehicles, we want to highlight the methodological "problem" that Bateson and Mead referred to as "camera consciousness." For their visual analysis of Balinese Character, Bateson and Mead took "about 25,000 Leica stills, and about 22,000 feet of 16 mm film" over a two-year period,[5] a dataset that could easily be described today as "big data," even if their methods of analysis were wholly qualitative. Of the 759 photographs chosen for their book, they note that only eight could be considered to be "posed." In this sense, Bateson and Mead explain quite explicitly, the factors that "contributed to diminish camera consciousness in our subjects."[6] They were well aware of the powerful effect of their presence on how people went about their everyday activities, and they were sensitive to the social impact of their cameras. Central to their technique was "the fact that we never asked to take pictures," and associated with this element of surprise was the use of "an angular viewfinder for when the subject might be expected to dislike being photographed at that particular moment."[7] They wore "two cameras day in day out so that the photographer himself ceased to be camera conscious."[8] With such a large dataset and habitual use of cameras, "it is almost impossible to maintain camera consciousness after the first dozen shots."[9]

In Bateson and Mead's attempt to produce new anthropological knowledge from their visual dataset, they encounter two parallel components to camera consciousness concerning, on the one hand, the camera and its operator and, on the other hand, the photographed (human) subject (Figure 2.1). This could be understood equally today in the tension around the dual sense of anxiety and augmentation in the relationship we have with cameras, in their potential for re-visioning the world and in the exchange and analysis of images as visual data. In fact, camera consciousness foreshadows something of the contemporary discomfort in the dual-use inherent in AI technologies that have begun to automate machine attention and action. For every development that automates vision or manages the use and distribution of digital images, there is a potential application for surveillance, governance or terror. So, it almost goes without saying that the components and the implications of camera consciousness have greatly expanded.

Camera consciousness describes an often-vexed emotional relationship with personal and impersonal cameras that has heightened with their proliferation and their embedded or associated data technologies. Points of view have become vastly expanded and often interconnected through networked visual systems or via visual social media platforms. Digital images carry a new operational and communicative power laden with information that can be newly exploited. Because of this, cameras have become "intelligent" (perhaps in some sense "conscious") through their capacity for machine vision, deep learning and real-time image processing aiding decision-making. Those same cameras can

FIGURE 2.1   Margaret Mead and Gregory Bateson working in the mosquito room, Tambunam, 1938. Gelatin silver print.

Source: Library of Congress. Copyright Bateson Ideas Group.

generate image-based logistical systems and, as we will show, bring into play a new race to commercialize and secure visual and sensory data infrastructures, which positions cameras as a key technology of the fourth industrial revolution and a functioning Internet of Things (IoT).[10] The automation of vision in seeing machines represents a long sought-after pinnacle of artificial intelligence – the pairing of cameras and judgment. But there are human and social implications. Ultimately, visibility remains as it was in Bateson and Mead's Bali fieldwork: an unevenly distributed commodity within the digital economy and experienced as the site or territory of social, political and economic struggle.

In this book, it is our contention that the "problem" of camera consciousness in Bateson and Mead's early visual anthropology points to a far more significant set of issues that can help us to frame and understand subsequent camera technologies and our still ambivalent countenance before and with cameras in the age of mobile media and autonomous seeing machines. Camera consciousness flags something important about the shifting locus of control and agency in the increasingly automated digital landscape. For one thing, Bateson and Mead brought to their work in Bali not just a camera lens with a nifty viewfinder extension but also echoes of a colonial perspective. We can only speculate, but it may be safe to say that their Balinese "subjects" were undergoing their own attempt to negotiate the technological intervention those cameras signaled, alongside the social interference of the researchers that accompanied them. The two are not easily separated.

Camera consciousness is an underacknowledged concept with a history that has spanned the development of cameras and cinematography for more than a century now. In early photography, it can be found in the newly "magical" (or scientific) properties of the camera as an apparatus with the ability to "write light" and also in the patience of a portrait sitter who had to pose perfectly still to accommodate long exposure times or in the slow framing and lighting requirements for recording a landscape picture. Similarly, the actors in the early cinema houses had to adjust from performing before people on a stage to addressing an imagined audience through the camera while finding a mode of address that ignored its presence within the scene.[11] We probably underestimate the magnitude of this transition. However, those who now take the common step of self-projection to turn their mobile camera on themselves and distribute selfies online may feel these relational negotiations more directly. The sometimes-debilitating self-consciousness and self-awareness that happen when acting before a camera were originally referred to as camera consciousness, a "disorder" that could be cured through mental training. But the consciousness of the *presence* of cameras and the consciousness *of* cameras is a distinction that's easily blurred.

Cameras have become more active or agentic,[12] and our modes of visual communication more diverse than ever. The value of visual data – the image content along with an image file's digital properties, marking the time, location and many other pieces of trace data – becomes most apparent in automated and mobile technologies such as drones, face recognition, mobile phones and driverless cars. But that value is dispersed across many more areas of social life and is put to analytical use in a range of sciences (geology, archaeology, engineering, medical imaging, criminology, astronomy). We see in these technologies and contexts evidence of a quickening of the datafication of the visual and the rising value of everyday visual data. New developments in camera technologies and image processing enable Bateson and Mead's visual anthropological techniques to be turned in every direction in the name of solving problems of visual analysis at scale.

It is time to look at cameras and images differently – to understand cameras as lively assemblages invested with intensity and disruptive force as well as augmentative power. By considering images as image-making, or *imaging*, as bursting with the new value of visual data, we can begin to understand the new functionality they offer – they are dense constellations of potential information, linkable, searchable, analyzable, networked, structured, layered digital objects. The new camera consciousness responds, often imperceptibly, to this barely visible element of visual data tied up in multiplying forms of metadata and analyzable pixel arrays. It points to the awareness of, awareness with, and awareness embedded in networked and intelligent cameras. There's an urgency to these developments, a buzz in the technology press and research journals exploring new applications for image recognition and processing, mobile sensing and mapping.

The point of taking camera consciousness as a framework for analysis of these new technologies is to help pinpoint the locus of control or agency at each *contested* point of their application and development. Camera consciousness offers a kind of intuitive and affective test for spotting the changing social relations surrounding new image processing technologies. This is a test that Google Glass, for example, already failed without anyone really knowing ahead of time why. With equal measure of interest, anticipation and outright derision, Google Glass – essentially always-on camera and visual augmentation-enabled glasses – failed, more socially than technically, to overcome the anxiety associated with camera consciousness. But more than this, if the majority of people are to accept the widespread introduction of driverless cars, or pilotless planes, there has to be a critical level of trust in their competency, their infallible adaptive awareness. In these technologies, we are waiting for evidence of a camera consciousness that demonstrates an ability to act with an even greater intelligence than human operators. The social implications of the shift that follows this tipping point are significant. It is at this point that the social becomes envisioned differently, and visual data becomes an infinitely important new commodity.

The case studies we offer are face recognition technology, drone vision, mobile and locative imaging and driverless cars. These case studies are tied together by the relative autonomy of their respective "vehicles" for dispersing points of view. Their cameras are usually on the move and embedded in the world. To work at all, they have to cross a threshold of *awareness*. And if they do so, they pave the way for integrating the increasingly complex and intelligent camera and image processing technologies they deploy into social settings and into our everyday lives. The subsequent applications and implications are broad reaching. Our focus lies in understanding the disruptions to existing ecosystems, to urban environments, everyday social practices and visual publics, and the new social relations that camera conscious seeing machines bring into being.

Vision, Visibility and Vitality

We take seeing for granted as a particularly human or at least animal activity. And yet, the history of media has been banally entwined with practices and technologies of visualization, visual communication and visual literacy, often carved up and understood in genres and spheres of image-making practice. John Berger made the point that in more rigid aesthetic categories of the image, such as a painting, portrait, landscape, photograph or video, "what is forgotten – like all questions in a positivist culture – is the meaning and enigma of visibility itself."[13] Berger's influential books *Ways of Seeing* (1972) and *About Looking* (1980) were published at a time of seeming transition, when the canons of modern visual culture were finally shaken off their perch by personal cameras, television, cinema and advertising, making visible the ideological underpinnings of visuality – of seeing and being seen. What scholars such as Berger, Susan Sontag, Roland Barthes and others have taught us is that "the visual in all its complexity can be broken down for the purpose of analysis and criticism, but the essential components of vision will remain enigmatic," as Ron Burnett explains.[14] Vantage point, attention, comprehension, for instance, are all essential to the active work of vision, and hence to its automation and computational use among digital platforms and applications.

We are experiencing another point of transition in visuality and visibility, not perhaps a new "scopic regime"[15] but rather a newly "emerging power to visualize," as Vilém Flusser puts it.[16] Where cultural anxieties once surrounded the representational and indexical developments in photography, film and television and their popular cultures, anxiety now often follows the proliferation of cameras, their ability to see, recognize and process visual data all by themselves. Or it entails the power relations they make apparent. Vision and visibility are augmented by computer processing and robotics, mobilized, processed and dispersed through an increasing range of networked media devices. Every smartphone sees or senses the world, creates and circulates digital images and, in this sense, reshapes or affects our visual publics. If public spaces are those in which we are visible to others, they become confused by every sensory intrusion into those remaining spaces that disrupt that visible access in the name of privacy. So, the enigma of visibility turns to generalized anxiety. In the contemporary tradition of surveillance studies, there are many accounts of the disruption of the boundaries of public and private brought about by the proliferation of CCTV and other always-on digital imaging systems.[17] The narrative of our time is that progressively, technologies of visibility, to use Andrea Brighenti's phrase,[18] are reconfiguring publicness, and the notion of some protected private sphere is dissolving. However, these concerns should stand only as a starting point to understanding our new domains for visual media and the implications of the new camera consciousness.

To appreciate the contemporary value of visual data and the power inherent in the datafication of the visual, the concept of camera consciousness introduced earlier can be broadened through reference to a particular line of social theory that, among other goals, has sought to rethink the materiality and agency of objects and images and the composition or assemblage of machines. This line of social theory spans the work of Henri Bergson, C.S. Peirce, William James and Gilles Deleuze through to recent accounts of actor network theory (ANT), vital materialism and the nonhuman, in the work of Jane Bennett, Diana Coole and Samantha Frost, Karen Barad and other feminist proponents of new materialism.

In the philosophy of Deleuze and Bergson, an image is as much a part of matter as any other thing. This unorthodox understanding of an image can be difficult to accept given the way we commonly talk about images as immaterial – as representations *of* things, as fundamentally opposed to things themselves. But, the materiality of the image as a component of media and communication is becoming more pressing in the digital age and in the age of ubiquitous mobile imaging and image sharing.[19] In Deleuze's philosophy, any image refracts or slices the world in different ways, sometimes emphasizing movement, sometimes perception, affection or time (duration).[20] As a starting point, this understanding of the world as image, and image as matter, movement, affect, impulse, time and so on, is critical in deprioritizing human perception in vision and hence accounting for the technological developments we explore in this book. Does a driverless car *see*, or does it use and compute a massive set of refracted slices of the world we might refer to as visual data? How do these "images" inform steering, acceleration or breaking? How do they distinguish between a pedestrian and a roadworks barrier, let alone make decisions about how to navigate them?

In his brief discussion of camera consciousness, Deleuze points to the "semi-subjectivity" of the mid-twentieth century cinematic image, an image that splits the point of view or vision of the world in a transformative process. In a broad sense, Deleuze's concept of camera consciousness is connected with that of Bateson and Mead, as already described. And, it can be reflected more generally in the dual countenance of contemporary digital visual media practices: in the feeling that makes us pause, hurry past or turn away when a camera appears before us in a crowded street; or, alternatively, to pose, to edit, frame, filter, caption and share the images we make daily through camera phone use. The camera, the image, our body all have some sense of agency in this process of imaging, with any number of consequences. As Jane Bennett puts it in relation to what she describes as the "enchantment" of some things, their "thingly power": "organic and inorganic bodies, natural and cultural objects (these distinctions are not particularly salient here) all are affective"[21] – all act on the world and bring into being new relationships.

There should be little doubt that both the digital camera (or drone or mobile or driverless car camera) and the image it is able to generate possess their own "vitality" or "thing-power."[22] Even looking back at the tape covering the small aperture at the top of a laptop illustrates the ever-expanding technical infrastructure "making the camera felt" forming part of our contemporary mediated ecosystem.[23] For Deleuze, camera consciousness becomes visible in what he calls the perception image, or in the imaging of perception. He isolates examples such as the experimental cinema of Dziga Vertov, where the "semi-subjectivity" of the camera becomes apparent, and in the break with the sensory motor schema in post-war Italian neorealism producing "pure optical and sound situations" that exceed the limits of action and narrative.[24] Camera consciousness splits the point of view or vision of the world, of both characters and cameras, in a transformative process:

> We are no longer faced with subjective objective images; we are caught in a correlation between a perception-image and a camera-consciousness which transforms it (the question of knowing whether the image was objective or subjective is no longer raised).[25]

In the process, the relative autonomy of the camera becomes apparent. The camera enters its own "mental connections," and so participates in "questioning, responding, objecting, provoking" and otherwise acting.[26]

To understand the materiality of camera technologies and digital images, or the affective dimension of camera consciousness, it is useful to apply recent approaches to the nonhuman,[27] and to consider machine-body assemblages, or the materiality of media and communication[28] or the vitality and "lifeness" of mediation as a *process*.[29] Rather than embedding analysis in a critical account of the surveillance potential of autonomous visual technologies, we explore their vitality within emerging new media and urban ecologies. It is better to try to understand how new camera technologies affect and reconfigure the territory of the visible. Jane Bennett, for instance, draws attention to the enchantments of objects and builds on approaches to technology that foreground the "agentic" capacity of nonhuman objects within any assemblage or network, as elaborated through the actor network theory of Latour, Callon and others.[30] When face recognition, for instance, becomes a tool for unlocking a laptop or opening a door, the locus of control shifts from a person with a key or passcode to a camera conscious system acting as gate keeper and security guard. The visual data of a face becomes the codified key that opens a door or forces it to lock shut, but it is the camera conscious system that takes control.[31]

Considering the digital, mobile or vehicle-driven camera and associated image recognition and processing technologies through these frameworks helps to better understand their emergent value. We have moved beyond the age of the human-centric image – photographic, cinematographic or digital – to find ourselves in a time of ceaseless production, manipulation and circulation of analytic modes of social imaging. The new role of the camera and digital image in transforming the social should not be underestimated. Trevor Paglen has described this transformation by emphasizing the shift from the representational and human-oriented qualities of images to their operational and activating functions:

> Human visual culture has become a special case of vision, an exception to the rule. The overwhelming majority of images are now made by machines for other machines, with humans rarely in the loop.[32]

Camera consciousness also suggests the need for a kind of pragmatics, or to call on philosopher William James, a necessary rethinking of consciousness in the form of a "radical empiricism." James sought to move beyond the transcendentalism of Kant and the dualism of thought and thing, object and subject, which, in the context of media and Internet cultures, popularizes the unhelpful and untenable notion of a fundamental separation between a digital and physical self, an online and offline world, a world with and without technology and digital traces. In James's radical empiricism, "consciousness connotes a kind of external relation and does not denote a special stuff or way of being."[33] The "stuff" of ubiquitous digital imaging is bound up in the particularity of experiences, their plurality: "they not only are, but are known," we are aware of their qualities, and they act as the content of knowing; but this awareness is itself an experience of relations, "of different degrees of intimacy."[34] For James:

> Our fields of experience have no more definite boundaries than have our fields of view. Both are fringed forever by a *more* that continuously develops, and that continuously supersedes them as life proceeds.[35]

The Social Lives of Cameras

Cameras began to take on new social roles when networked infrastructures improved in the first decades of the Internet, and web-enabled cameras unshackled "tele-vision" from the exclusive domain of broadcast media companies.[36] Webcams made intimate personal lives public and political in the late 1990s and early 2000s through new modes of distributed telepresence that offered a tethered but nonetheless altered transformation of visibility.[37] As Terri Senft showed, webcams were great equalizers and strange new socialization tools able to capitalize on emerging forms of networked intimacy; their effects are now extended through the broad array of video streaming and sharing tools among social media platforms. Indeed, Facebook's current iteration of video connectivity, called *Portal*, aims to make use of an automated laser-guided camera so as to detach the camera's point of view and presumably enhance – further naturalize – real-time visual communication. Like Bateson and Mead's Balinese research subjects, a transferal of camera consciousness is deemed central to their embeddedness in our everyday lives.

Only half a decade on, the selfies moment of 2013 seems in some ways almost a distant memory. Has the time of selfies passed? Perhaps not passed so much as become more opaquely embedded in the regimes of face-data collection and processing – both a communication and machine vision infrastructure. The early 2010s saw the spread of smartphones, front-facing cameras and image-sharing apps like Instagram (2010), Snapchat (2011) and Instagram Stories (2016). In a short time – but with a long trajectory of associated cultural practices – we have become more competent with communicating or socializing through selfies and the language of facial gestures. The selfie moment involved a "political convergence of the object and subject of photographic practice."[38] What we refer to as digital participation, or digital citizenship, implies a compliance with this convergence. Smart cameras catch us in the nexus between object and subject of imaging. "A selfie is a way of speaking and an object to which actors (both human and nonhuman) respond,"[39] where the value of the face image goes well beyond its role in person-to-person communication to include every operation made possible by automated image processing technologies.

And, as we discuss at the end of this book, semiotics also has a renewed role to play in the development of new visual literacies. Within supervised machine learning, automated machine vision systems are able to both interpret and produce elements of the visible world as inputs and outputs through recognizable semiotic processes.

Where does all of this leave us? If we lose sight of the precarious balance of agency in new camera technologies, or the equivocal nature of camera consciousness, there is little hope in discovering their social good. This is about looking at but also *beyond* the technologies or devices themselves. We experience *relationships* with and through cameras, with and through images and with and through selfies, but in ways that always feels like there's *more*. As with the framework and methods of Actor Network Theory (ANT), radical empiricism doesn't jump to the thing alone to situate analysis with a genre of media content – the drone, the smartphone device or driverless car *in themselves*. Both approaches – ANT and radical empiricism – also consider the relations and processes in between, their conjunctions and disjunctions, and so foreground their forcefulness and the depth of their impact within media and social ecologies.

Camera consciousness is our conceptual tool for expanding digital and data literacies, to better account for the value and potential exploitations or fallibilities of smart cameras and digital images. We want it to imply an awareness *of* cameras, alongside an understanding of the full implications of *camera awareness* built on the readability, value and operability of visual data. Camera consciousness can be expressed as an awareness of the various *perceptual relations* described earlier and throughout the book, an awareness of seeing – or its augmentation – and being seen, which we experience as no less forceful than the *content* of perception, of images or of a face-to-face connection.

In different ways, acts of imaging or visualization and circulation extend connectivity and establish new kinds of relational experience. This aligns with the expressive imperatives of social media as both a generalized and individualized field of view. And, it lays the groundwork for understanding the vitality and liveliness of the devices and vehicles underpinned by machine vision and visual processing. Whether explicitly described or inherent, these theories of image and camera power are embedded within the broad set of technologies, systems, processes, discourses and social circumstances that have surrounded autonomous devices like drones and driverless cars or systems enabled by face recognition and the unfurling potential of mobile camera devices. These apparatuses of visualization carve out new ways of being urban, of being social or commercial, new ways of seeing and communicating.

The Value of Visual Data

There is an emergent and by no means settled value in the camera technologies that drive a growing range of autonomous seeing machines. The popular conflation of camera conscious technologies with surveillance or with "autonomous weapons systems" need not be absolute.[40] Bishop and Parikka offer the example of the Planetary Skin Institute (PSI – NASA and Cisco Systems), among other global intelligent remote sensing systems that automate environmental surveillance to emphasize the positive uses to which artificial intelligence technologies can be put. As the PSI's promotional video explains, "we can't manage what we can't measure."[41] In emerging, unfinished technologies such as personal mobile media devices, drones and autonomous or driverless vehicles, we are witnessing a further expansion of our understanding of media and a reconfiguration of the material and social environments in which they operate. As icons of the "sensor society," to use Mark Andrejevic's terminology,[42] these kinds of technologies invite us to consider again what counts as media and how mediation takes place and affects us. We offer a way of making sense of the increasing range of computer vision technologies, or technologies of visibility, and the vehicles that make them mobile, through an understanding of the new camera consciousness they bring into being and reflect.

To step back a little, what is the metadata stored in an image file? Every digital image contains rich possibilities and is packed with descriptive, technical and administrative metadata. Putting aside their uses and abilities to convey meaning, to denote and connote or to signify, any digital image can be described as a *bundle of information* that includes the pixel data, the aesthetic components we see, along with dense and expansive layers of data and textual overlays. These elements of a digital image are technically referred to as metadata, but we simplify here as data to position the digital image itself as an item of rich informational depth and potential. An image file for a digital photo can include *descriptive data* – Dublin core categories such as creator, date and time, format, identifier, rights, subject, title, description, longitude and latitude, etc. – and *administrative data* – a form of technical metadata that comes automatically, for example, with a digital camera's use of the Exchangeable image file format, or Exif, which includes information associated with the camera device and image properties, such as image orientation, resolution, flash on or off, focal plane, ISO speed, exposure mode or bias and color space, amongst other data points.[43] Structural metadata is included in video files to signal the relations between image and audio, order of play and so on.[44] Lidar – the visual system used for measuring distance and spatial arrangements – creates a data arrangement in the form of a 3D point cloud dense with actionable information. In fact, imaging developments such as these are all about generating and extending the functionality of visual data.

It is worth asking how the camera and image came to acquire such capacity for producing and processing visual data. Not all that long after Bateson and Mead were using novel camera techniques in Bali, pioneering computer science engineer and mathematician Azriel Rosenfeld[45] turned his attention to digital image processing techniques. His 1969 book, *Picture Processing by Computer*, was an early guide to a decades-long pursuit of the science and possible computational applications for reading and manipulating digital images. Work on image processing accelerated just two decades after Bateson and Mead's visual anthropology in Bali, with MIT's Bell Laboratories as the central engine room in the 1960s. Although the contemporary applications of machine vision and image analysis seem to have appeared suddenly, the work of Rosenfeld and many others have gradually enabled developments in satellite imaging, medical imaging, videophone, photo enhancement and manipulation and face recognition, among other applications and fields of knowledge.

Image processing involves the application of algorithms to digital images to achieve everything from projection to classification, pattern recognition, image corrections, filters and other forms of processing and analysis. These are the foundational technologies that enable digital cameras to capture and form images, add metadata, and undertake all the intelligent processing, manipulations and analytical applications that may follow. Other sensory arrays and location and spatial awareness systems, like motion sensors, accelerometers or lidar (light detection and ranging), combine to extend the terrain of visibility, generating dynamic environmental mapping, rendering even the most intimate, personal spaces operational in new ways.

The growing utility of digital images lies in their connections with other forms of *use data* or data exhaust, the data left behind by all digital and online activities through cookies, log files and temp files for instance.[46] The rich information imbedded within digital images enables what Nadav Hochman refers to as "the social image" in the move from databases to virtual "real-time processing of continuous data flows from geographically distributed sources" in the form of a "data stream" – the rapid, algorithmic curation and flow of images for social platforms and other cloud-based technologies.[47] This is where the networked capacity of digital images, camera-oriented devices and visual communication comes into play. A driverless car both maps its pathway visually and responds to an existing and continuously updated wealth of already mapped and processed data. It produces visual data and functions in relation to data already produced and processed. The connectedness of digital and online acts with cameras and images bridges the world we (and autonomous objects) see and traverse with other forms of searchable, traceable online activity. Visual data plays a particularly interesting role.

To take just one field of applications for visual data as an example, remote sensing for archaeological and geo-mapping relies on comprehensive imaging and intelligent data processing and analysis techniques. Its goal is to reveal useful geological or archaeological information not available to the human eye. Remote sensing makes use of satellite or high-altitude photography and photogrammetry (the science of taking measurement from photographs) through often autonomously acting vehicles and cameras to visualize, analyze and categorize objects or features on earth. Camera, vehicle and data processing software combine to produce new forms of visual information. As infrastructures of a global vision, satellites, for instance, proliferated throughout the second half of the twentieth century, reinventing global maps and reshaping the image of earth. These are, as Laura Kurgan has detailed, "the technologies that produce global imagery and that both necessitate and facilitate the interpretation of images at once measurable and digital, uncentred and ambiguous, yet comprehensive and authoritative."[48]

That is, while satellites form an unblinking eye from above, their vision stitched together to "reveal" the terrain of the entire planet, they are *activated* through their capacity to compute the social data in what they see – the built environment, the movement of troops or vehicles in war zones, the spread of cities across continents, the gradual shrinking of the polar icecap. The Geoscape project, for example, conducted by the company PSMA Australia used AI and machine learning in this way to operationalize satellite images to provide a 3D dataset of the 15,243,669 buildings across the geographically vast landscape that is Australia.[49]

Historically, every new apparatus of visibility has been closely tied to the information it is able to produce. Take the hot air balloon, for example. In his book *Seeing the Social: Selected Visibility Technologies* (2010), Harry Freemantle includes the balloon in his exploration of the development of "visibility technologies," alongside perspective lenses, the camera obscura, the lithograph, diorama and photography, offering a similar perspective to Crary's *Techniques of the Observer*.[50] Experiments in balloon flight such as those of Jacques Étienne and Joseph Montgolfier in France in 1783 represented not only a new vehicle of transportation but also a shift in visibility. The balloon created a *platform* for a new vision of the city. Freemantle cites the French meteorologist Bertholon, an early pioneer and enthusiast:

> I have no doubt that the cause of science will be helped by the superior method of observation offered by balloon flight. In order to record accurately all that is observed in flight I urge the use of a camera obscura in order to transcribe what is seen.[51]

In some ways the step is subtle from this feat of engineering and scientific observation to Bateson and Mead's visual anthropology of Bali, and to geothermal mapping through more powerful satellite or drone cameras. A whole evolving ecology of visibility technologies follow: including personal mobile devices loaded with sensors and autonomous vehicles seeing and sensing their way around urban environments while feeding new visual and geospatial data into already dynamic maps.

What's happening here, behind the scenes, is the emergence of new infrastructural platforms that bring together camera technologies, automated digital imaging and cloud-based visual data storage (for real-time processing) with new possibilities for computation. To adapt Benjamin Bratton's concept for the material conditions of computation in the notion of "the stack," we are facing the development of powerful new *platforms* that connect and operationalize visual data, "pull things together into temporary higher-order aggregations and, in principle, add value both to what is brought into the platform and to the platform itself."[52]

When camera technologies become computational by default, standardized through data structures and metadata conventions, and as they become in the process intelligent, we see "the capitalized translation of interactions into data and data into interactions." This translation forms the basis of new platforms, new infrastructure or new utilities that build applications with visual data. They could take the form of Google Street View, or China's ambitious automated social credit system. Each brings into being new social relations that may provoke a buzzing anxiety, or offer immense possibility in reshaping how we live and interact with our increasingly digital and connected environment. To know more about these connections is to have the tools for negotiating policies, regulation, norms, ethical protocols and everyday practices in response.

Bateson and Mead saw the value of a particular kind of visual data that overcame camera consciousness to offer scientific insights into the "character" of a people. But a broader concept of camera consciousness is needed now to realize the visual knowledge generated by distributed, mobile, aerial and autonomous cameras and our modes of acting with them. We would suggest that camera consciousness is the paradigm through which the value of visual data makes sense – both in its anxieties and augmentations. Developing a keener sense and understanding of the new camera consciousness at play here involves ongoing social research and technical, cultural, experiential and political-economic elaboration and analysis. We need to know more about the technologies and their capabilities (and limitations), as well as the corporations, start-ups and institutions working to build the platforms that are shaping our camera conscious future.

What connects the technologies of visibility that we explore in the following chapters is the altered set of social relations they produce. Each oscillates in the public imagination as things of recoil and anxiety and systems of great potential augmentation or extension. Automation, machine learning and image processing are key components of the camera conscious assemblage, but so are the practices and activities of those who benefit or are subjected to the dilemmas these vision-enabled technologies create. We are only just beginning to see the potential in visual data, but as a media and communications project the goal needs to be how this potential can be mobilized and democratized beyond the profit motives of technology corporations.

Notes

1. Gregory Bateson and Margaret Mead, *Balinese Character: A Photographic Analysis* (New York: New York Academy of Sciences, 1942).
2. Ira Jacknis, "Margaret Mead and Gregory Bateson in Bali: Their Use of Photography and Film," *Cultural Anthropology* 3, no. 2 (1988): 160–177.
3. Bateson and Mead, *Balinese Character*, xii.
4. Ibid.
5. Ibid., 49.
6. Ibid.
7. Ibid.
8. Ibid.
9. Ibid.
10. Mercedes Bunz and Graham Meikle, *The Internet of Things* (Cambridge: Polity Press, 2017).
11. An interesting account of camera consciousness is offered by the actor William Shatner. In an interview available on YouTube, Shatner responds to questions about his move from stage acting to television. He explains that he began to feel that the cameras were an audience. "Cameras were very large and they had a hot tube in it, and they had a fan in it; and the fan . . . made a little whirring sound as it cooled the innards of the camera. For all that you might know, it was breathing. . . . So I thought of the camera as alive. It was like . . . if you stood beside it to make an entrance, the thing kind of purred at you. . . . Well, you're aware and not aware. If you're aware then you shouldn't be aware . . . it's a very Zen thing . . . a kind of mystical position": See: "William Shatner on Camera Consciousness", YouTube video, 8:06. "shatnerbiter," February 10, 2009: www.youtube.com/watch?v=TFVGgP4oxAA&t=5s.
12. Joanna Zylinska, *Nonhuman Photography* (Cambridge, MA: MIT Press, 2017).
13. John Berger, *About Looking* (New York: Pantheon, 1980), 41. *About Looking* followed Berger's highly influential book *Ways of Seeing* (London: Penguin, 1972).
14. Ron Burnett, *How Images Think* (Cambridge, MA: MIT Press, 2004), 11.
15. Berger, *Ways of Seeing*; Hal Foster, ed., *Vision and Visuality* (Seattle: Bay Press, 1988); Jonathan Crary, *Techniques of the Observer: Vision and Modernity in the Nineteenth Century* (Cambridge, MA: MIT Press, 1990).
16. Vilém Flusser, *Into the Universe of Technical Images* (Minneapolis: University of Minnesota Press, 2011), 36.
17. See, for example: David Lyon, *Surveillance After Snowden* (London: Polity, 2015).
18. Andrea Mubi Brighenti, *Visibility in Social Theory and Social Research* (Houndmills, Basingstoke, Hampshire: Palgrave Macmillan, 2010).
19. See, for example: Connor Graham, Eric Laurier, Vincent O'Brien, and Mark Rouncefield, "New Visual Technologies: Shifting Boundaries, Shared Moments," *Visual Studies* 26, no. 2 (2011): 87–91; Martin Hand, *Ubiquitous Photography* (Cambridge: Polity Press, 2012).
20. Gilles Deleuze, *Cinema I: The Movement-Image*, trans. Hugh Tomlinson and Barbara Habberjam (Minneapolis, MN: University of Minnesota Press, 1986); Gilles Deleuze, *Cinema II: The Time-Image*, trans. Hugh Tomlinson and Barbara Habberjam (Minneapolis, MN: University of Minnesota Press, 1989).

21. Jane Bennett, *Vibrant Matter: A Political Ecology of Things* (Durham, NC: Duke University Press, 2009), xii.

22. Ibid., 2.

23. Deleuze, *Cinema I*, 74; Patricia Pisters, *The Matrix of Visual Culture: Working with Deleuze in Film Theory* (Stanford, CA: Stanford University Press, 2003).

24. Spencer Shaw, *Film Consciousness: From Phenomenology to Deleuze* (Jefferson, NC: McFarland, 2008), 161; Valentine Moulard-Leonard, *Bergson-Deleuze Encounters: Transcendental Experience and the Thought of the Virtual* (Albany, NY: State University of New York Press, 2008), 110.

25. Deleuze, *Cinema I*, 74.

26. Deleuze, *Cinema II*, 23.

27. Richard Grusin, ed., *The Nonhuman Turn* (Minneapolis: University of Minnesota Press, 2015).

28. Tarleton Gillespie, Pablo J. Boczkowski, and Kirsten A. Foot, eds., *Media Technologies: Essays on Communication, Materiality, and Society* (Cambridge, MA: MIT Press, 2013); Jeremy Packer and Stephen B. Crofts Wiley, eds., *Communication Matters: Materialist Approaches to Media, Mobility and Networks* (London: Routledge, 2012).

29. Sarah Kember and Joanna Zylinska, *Life After New Media: Mediation as a Vital Process* (Cambridge, MA: MIT Press, 2012).

30. Bennett, *Vibrant Matter*.

31. This is an account famously foreseen by Gilles Deleuze in his 1992 essay: Gilles Deleuze, "Postscript on the Societies of Control," *October* 59, no. 1 (1992): 3–7.

32. Trevor Paglen, "Invisible Images (Your Pictures Are Looking at You)," *The New Inquiry*, December 8, 2016, https://thenewinquiry.com/invisible-images-your-pictures-are-looking-at-you/.

33. William James, *Essays in Radical Empiricism* (New York: Longmans Green and Co., 1912), 17.

34. Ibid., 27.

35. Ibid., 38 – emphasis added.

36. Michael Wesch, "YouTube and You: Experiences of Self-Awareness in the Context Collapse of the Recording Webcam," *Explorations in Media Ecology* 8, no. 2 (2009): 19–34; Daniel Miller and Jolynna Sinanan, *Webcam* (New York: John Wiley & Sons, 2014); Jean Burgess and Joshua Green, *YouTube: Online Video and Participatory Culture*, 2nd ed. (London: Polity Press, 2018).

37. Theresa M. Senft, *Camgirls: Celebrity and Community in the Age of Social Networks* (New York: Peter Lang, 2008).

38. Edgar Gómez Cruz and Helen Thornham, "Selfies Beyond Self-representation: The (Theoretical) F(r)ictions of a Practice," *Journal of Aesthetics & Culture* 7, no. 1 (2015), https://doi.org/10.3402/jac.v7.28073.

39. Theresa Senft and Nancy K. Baym, "Selfies Introduction – What Does the Selfie Say? Investigating a Global Phenomenon," *International Journal of Communication* 9 (2015): 1589.

40. See, for example: Ryan Bishop and Jussi Parikka, "The Autonomous Killing Systems of the Future Are Already Here, They're Just Not Necessarily Weapons – Yet," *The Conversation*, August 4, 2015, https://theconversation.com/the-autonomous-killing-systems-of-the-future-are-already-here-theyre-just-not-necessarily-weapons-yet-45453.

41. "Planetary Skin Institute – Global Nervous System," YouTube video, 3:43. "Kostas," November 28, 2010. https://youtu.be/K3K990kVLS0.

42. Mark Andrejevic and Mark Burdon, "Defining the Sensor Society," *Television and New Media* 16, no. 1 (2015): 19–36.

43. Jeffrey Pomerantz, *Metadata* (Cambridge, MA: MIT Press, 2015).

44. Ibid.

45. Azriel Rosenfeld, *Picture Processing by Computer* (New York: Academic Press, 1969).
46. Ibid.; Alex Williams, "The Power of Data Exhaust," *TechCrunch*, May 26, 2013, http://techcrunch.com/2013/05/26/the-power-of-data-exhaust/.
47. Nadav Hochman, "The Social Image," *Big Data & Society*, July–September (2014): 1–15.
48. Laura Kurgan, *Close Up at a Distance: Mapping, Technology and Politics* (New York: Zone Books, 2016), 13.
49. Jim Malo, "All of Australia's 15.2 Million Buildings Have Been Mapped," *The Age*, October 30, 2018, https://bit.ly/2qoo9DT.
50. Harry Freemantle, *Seeing the Social: Selected Visibility Technologies* (Fremantle: Vivid Publishing, 2010).
51. Quoted in Barbara Maria Stafford, *Voyage into Substance: Art, Science, Nature, and the Illustrated Travel Account, 1760–1840* (Cambridge, MA: MIT Press, 1984), 486.
52. Benjamin H. Bratton, *The Stack: On Software and Sovereignty* (Cambridge, MA: MIT Press, 2016).

3

FACE VALUE

Let's begin with the face.

> The face, what a horror. It is naturally a lunar landscape, with its pores, planes, matts, bright colors, whiteness, and holes: there is no need for a close-up to make it inhuman; it is naturally a close-up, and naturally inhuman, a monstrous hood.[1]

To map the lunar landscape, from earth at least, you need a powerful lens able to achieve high definition digital imaging and a lot of computing power. This is the perfect metaphor for face recognition technology in all of the socio-technical contexts in which it is increasingly used. Deleuze and Guattari, quoted earlier, were writing in the late 1970s, and so close-up access to the lunar surface hadn't receded too far from the public imagination. It was also a time when the "science" of the face was beginning to ramp up. The Facial Action Coding System (FACS), which built on the 1969 work of Swedish anatomist Carl-Herman Hjortsjö, was set out formally by Paul Eckman and Wallace Friesen in 1978 and later updated in 2002 in collaboration with Joseph Hagar. This system and others developed in the decades since have been used to formally categorize facial features and expression by coding the gestalt patterns of a face's symmetry and muscular movements in a way that is perfectly suited for standardized and automated computational analysis.[2]

Deleuze and Guattari's theoretical account of the face in their chapter of *A Thousand Plateaus*, "Year Zero: Faciality," has been described as both "profound" and "open to interpretation" by Kelly Gates in her work on face recognition technology, *Our Biometric Future*.[3] Similarly, the history of developments in face recognition technology has followed a bumpy path. In Gates' words, "the automation of face recognition and expression analysis" has a technological trajectory that was motivated early on by a generalized "prospect of creating more intelligent machines," rather than any "immediate or well-defined needs."[4] While these social needs and applications are beginning to take shape, it is with significant controversy. What Deleuze and Guattari point to, and Gates later argues, is that, as a project of both anatomy and computer science, the face should not be considered in any straightforward way the "mirror of the mind," or simply there to be read. For Gates, the assumption that the face is a universal form perpetuates problems with the accuracy of automated face recognition systems and how they embed racial and gender bias.

The face, or to use Deleuze and Guattari's term, faciality, is *produced* historically, economically, politically and socially: "It is not the individuality of the face that counts but the efficacy of the ciphering,"[5] that is, the encoding of, or the system of categorization for, the collection of features that comprise a face in an effort to formally produce a face as readable. What counts then is also the utility of the machine-readable face – the identifications, tracking, permissions and sanctions, inclusions and exclusions. At the same time that Deleuze and Guattari were warning of the particular kinds of power that make use of the face, other social and cultural theorists were talking about the mythical layers of meaning associated, for instance, with the cinematic "face of Greta Garbo" in the writings of Roland Barthes.[6] The machine-readable face differs from Barthes' mythology of the face (as icon and as constellation of cultural meanings and significations). The face as biometric marker is an enabler of socio-technical interventions and applications in a way that contrasts with the affective ambiguity of face as cultural signifier.

Despite all the well-stated fears about the threat posed by AI-based face recognition systems, computer scientists have spent decades constructing machines capable of seeing the face in relation to a precise system of standardized signs. They have been busy producing the face as machine readable, and establishing the hardware, cloud storage and platform infrastructure to make face recognition a consumer-grade service. Perhaps more than any other automated vision technology, face recognition is mired in the ambivalence that flows from competing perspectives on faciality as they confront the brute force of computer science development and Big Tech investment, and State or commercial applications.

To account for the historical context as well as the social impact and likely future for face recognition technology, we draw together three areas of research and application to offer a perspective based in the ambivalent poles of the new camera consciousness. Of course, face recognition systems are most disconcerting and controversial when they are put to use in the service of governance, surveillance and control systems. This is why news of China's social credit system in western countries took off when it was revealed that a core component was the enormous expansion of public CCTV surveillance camera networks (with a reported target of more than 400 million cameras by 2020).[7] China has signaled its intention to find new applications for automated face recognition to achieve governance goals. However, as many commentators point out, most technology companies and governments around the world are hardly innocent in their attempt to develop and apply the same visual data–based governance practices. So, we also explore the commercial interests paving the way for the "platformization" of face recognition – particularly the involvement of Amazon (Rekognition), alongside the other big five most powerful technology corporations, as they invest heavily in face-data collection and analysis as "infrastructural platforms."[8]

Perhaps counterintuitively, our starting point in this chapter, however, is not with surveillance technology as such but with the widespread complicity that has made face images – and selfies – one of the more complicated objects of mobile digital media cultures. At the same time that selfies became relatively ubiquitous social media content and domesticated as everyday communicative practice, the value in the visual data of faces expanded exponentially. In fact, face recognition technology makes most sense and has the greatest impact and implications, in a time of near ubiquitous social media use. It comes to life in relation to the depth of personal investment in camera-based image sharing, as in the rise of Snapchat, Skype, Instagram and Facebook Messenger. It is underpinned by the digital traces associated with public and personal Internet use. Selfies, for instance, far from being the perfect negative trope of the narcissistic mobile and digital age, and *in addition* to the role they increasingly play as modes of interpersonal communication, present a powerful base layer of personal visual data to be mined, combined and operationalized for any number of intelligent and automated services based on individuals and populations.

The social contexts, implications and future trajectory for face recognition systems are tied to whether there is a willingness to lean-in to being before and with cameras. It is tied to the story of the value of digital faciality and facial communication. And it is bound up with the interests of Big Tech infrastructural platforms that advance, exploit and find new applications for face recognition systems, producing the foundations for the social benefit, the exploits or the forms of governance and social control that can follow.

Trading Faces

The selfie moment of 2013, the year of its official inclusion and definition in the Oxford English Dictionary, brought with it some incredulous and at times mocking revelations that faces had acquired a newly democratized kind of value. As many historians of photography, portraiture, family snapshots and mobile media researchers have noted, while the selfie might have been brought into mass use by webcams and the rear-facing cameras and screen feedback of smartphones, they have a long tradition particularly in western cultures.[9] The convergence of device, screen and camera features with social media platforms has been widely researched and debated. Here we focus mainly on the contested value of the visual data capture embedded in mobile self-imaging as this is manipulated by, and contributes to, developing machine vision systems and techniques. We are starting to see the growing importance of the face datasets captured through selfie apps for amassing a corpus to be used in training machine vision face recognition systems.

Researchers have characterized selfies as a significant form of networked camera-based self-representation and a substantive genre of visual expression and shareable digital objects.[10] They play a role in altering the conditions of personal visibility, public intimacy and relationships. In this sense, as Lasèn explains, selfies consist of gestural practices mostly making use of the expressive and communicative features of the face (or faciality) that engender highly relational and interactive representations and, thus, they also *act* as objects to which others can respond and react.[11] In other words, they are communicative devices that have an important social function. Selfies also involve a technical and infrastructural context. For Hess, they are not only exchangeable digital objects but are part of an assemblage built around "the relationship of device, its connected networks, and the material spaces it documents and the user's relationship with each of them."[12]

One of the most fascinating aspects of selfies as digital and cultural objects is their untethering from their producer through the interventions of filters, alterations and image processing. In order for Snapchat to generate its famous Dog Face filter, for instance, face recognition, tracking and image analysis had to be embedded within the image capture process. Around the end of 2015 and early 2016, a number of mobile phone apps began taking advantage of developments in machine learning algorithms and face image datasets to offer simple and entertaining face transformation tools. Snapchat acquired Obvious Engineering in June 2015, a start-up working on a three-dimensional face scanning app, before releasing Lens selfie-mode in February 2016 with seven animated special effects face-following filters. Face beautification apps were also widely popular in South Korea, Japan, China and other Asian countries from around 2016. Apps reading and dynamically manipulating face images took the "superficiality" but also functionality of selfies to a new level. A fair degree of public scorn and derision followed. Technology writer Stuart Dredge wrote in a *Guardian* article profiling a number of "face swap apps":

> Will 2016 go down in history as the year we failed to heed the warning about artificial intelligence overthrowing humanity one stone at a time, because we got distracted by swapping faces with our friends, our pets or nearby breasts?[13]

The point about pets and breasts refers to apps' reported misperception of closeups of animals and other body parts used to "trick" the recognition system.

Putting aside the important issue of misrecognition, error and bias for now, it's worth unpacking the involvement or even complicity of mobile app users in setting the scene for the development of face recognition technology.

For a brief moment, around two years after its 2017 launch, FaceApp, one of the many face modification apps, saw intense public interest around the world. In July 2019, the self-described "AI face editor" was being downloaded around 1.8 million times per day, with users circulating images widely across social media followed by news media interest in the app and all the attention.[14] The app's convincing ability to filter faces to make them look older or younger or alter gender laid bare the extra-human capacity of machine vision. Not only could the front-facing camera see us and show us to the world, it could see beyond us now to *reveal* a future self, a past or alternative self. Technologically, the aging and de-aging work through the same principles we use to estimate or "de-code" someone's age by how their face looks.[15] This involves identifying and manipulating the shape and contours of the skin around the eyes, smooth, tight, weathered, drooping or wrinkled skin (skin elasticity), the shape of the lips and mouth, chin line, hair color and texture, including eyebrows, and many other barely perceivable points of distinction. Once these facial elements are determined, encoded and predicted, the face image can be filtered or altered accordingly. Even fully artificial faces can be composed and constructed, but only because information about faces is now so extensive, and because the raw materials and building blocks, the databases of labeled face images are now so large.

With the intense spotlight on FaceApp's transformational vision abilities, concerns turned quickly to questions of control, ownership and secondary uses of users' face image data. That the developer of FaceApp, Wireless Lab,[16] was a Russian company that fed popular concerns and fears, and as quickly as the app rose to prominence, its aged and gender-swapped images disappeared from social circulation under a cloud of warnings. Technology and security experts were warning of risks on Twitter, people began commenting on Facebook posts from friends touting their altered faces, with links to articles detailing the potential problems of handing over face data at such scale to a Russian tech company. For example, a review on the Google Play Store ratings and reviews section reacts to the idea of deceitful phone and data access:

> I was using this app for some time. But today I just read about the data it collects from our phone and what the company does with it, WOW! . . . I know I accepted the terms (even without reading them . . . my bad, of course) but I'll NEVER install this app on any device.[17]

These reactions represent the surfacing of a camera consciousness as an awareness of the value of personal visual data on the part of many app users (but certainly not all). But there are two sides to the camera consciousness underpinning the tensions caused by the presence and social functioning of face recognition and AI in camera technologies. From the machine vision side, the augmented camera consciousness involved in these apps and in face recognition and analysis systems generally relates to the automated process of encoding and decoding faces as digital objects and calculations that allow for accurate manipulations, identification or awareness of expression. For those who stand before smart cameras, the question is not just about accuracy but also a different kind of awareness – of the altered visibility and simultaneous data collection, creation and processing taking place – and hence they pose the question of agency or control. Revelations of data breaches happen with such regular occurrence that they barely have any effect on users of apps and platforms.[18] The real question is where the value is placed in personal data and the consequences of ceding some of it for personal or social gain.

Both the enthusiastic ignorance that led to millions of users playing with visions of their face and circulating them among friends *and* much of the fear mongering about identity theft and troll farms miss the learning moment offered by FaceApp, Snapchat, Meitu and other AI-mediated selfie filters and apps. Historically, we are in a phase of "value capture." Essentially, value capture refers to the secondary value or payoff created in the process of investing in and building expensive infrastructure or services. In urban investments, for example, infrastructure projects like a new rail line or improved sewerage increase the adjacent land value, which can be captured directly through property taxes and rates or more complex public-private financing arrangements. Robert Gehl details something similar at work for early web 2.0 social media platforms, as users generate secondary value in the database through their interactions with posts.[19] For app developers investing time and energy in face recognition techniques, a cyclical value capture is built in as millions of users contribute their face image, captions and other data to an amassed dataset that in turn contributes to resourcing, training and perfecting its machine learning techniques.

In the race to develop face recognition systems and applications, technology companies with the greatest access to accurately labeled face image datasets possess the greatest advantage. Value capture can happen at a number of points in the chain of face data collection, processing and analysis. Face image training databases are the assets built laboriously for the purpose of establishing effective automated systems. For a machine to recognize faces, identify individuals or analyze face expression, it needs to have a working understanding of the end-goal. This is where training data comes into play. The bigger the dataset, and more accurately labeled, the more effective the face recognition and analysis system will become. The value capture formula is simple: machine vision allows the capture and coding of face images at scale, in turn expanding the dataset from which machine learning algorithms draws to more accurately qualify and improve the system's accuracy.

Both open access and private or commercial face image databases have been available for decades.[20] Datasets like the Cohn-Kanade AU (coded facial expression database), PIE Database CMU, Yale Face Database (A and B), MegaFace, CelebFaces and Faces in the Wild range in size, subject matter and the type of labeling used. Other datasets for machine vision are not restricted to faces, or incorporate faces and expressions among a wide range of objects. Kate Crawford and Trevor Paglen explore the immense Stanford University training dataset ImageNet.[21] With more than 14 million images labeled with over 20,000 categories, ImageNet continues to be a critical asset and infrastructure for machine vision, even if they are coming under scrutiny for acquiring face images without consent. Datasets also vary in how the images were collected (often celebrity face images are used because they are considered fair game for their public status). For this reason, older image sets are often smaller, numbering in the hundreds or thousands of images, but also because of the labor-intensive manual collection and annotation techniques involved in compiling them. More recently, with automated collection and access via social media platforms, apps (thanks to favorable terms of service agreements) and other means, and through automated labeling techniques, databases have grown to comprise millions of face images.

IBM has attracted attention and criticism for its sourcing of face image datasets in its efforts to develop a recognition and analysis systems. IBM's "Diversity in Faces" (DiF) dataset contains 1 million annotated images drawn, IBM says, from a 100 million strong creative commons face image dataset. IBM's stated goal was to "advance study of fairness in face recognition systems."[22] Most of the images were taken from photo hosting site Flickr without notifying their producers or subjects. While the images were sourced via creative commons licenses, IBM and others, like Microsoft's MSCeleb database of 10 million face images,[23] have attracted intense public scrutiny for not seeking consent to use the images for face recognition systems.[24] What perhaps seemed to the technology companies to be a valid way to establish better face recognition systems and informing academic and technology researchers is experienced differently by those whose face images are being used outside of their knowledge and control.

This tension in control over the face as a personal digital object is not a straightforward issue of property rights or privacy rights. There is public interest in improving the accuracy of face recognition systems and avoiding racial and gender biases. In their developer blog post introducing the DiF dataset, IBM argues that the problem with AI face recognition systems is not with the technology itself but with the training data. Training sets just have to be big and diverse enough "that the technology learns all the ways in which faces differ" and "to accurately recognize those differences in a variety of situations."[25] As Ellen Broad argues, diversity in training datasets is only part of the issue in combatting bias and inaccuracy. When developers construct detection and analysis algorithms, whether consciously or not, they embed fairness criteria to ensure that the systems adequately work for nontypical faces, racial and gender difference. Decisions have to be made about how to redress underrepresentation of minority faces, and how the algorithm treats those faces against what it has been programed to consider a typical or gold-standard face.

Before realizing the importance of diversity and the variety of settings, camera angles, skin tones and other complicating factors, researchers developing face recognition and analysis placed a lot of attention on the encoding and decoding techniques to map or mask faces accurately. Yet, approaches to coding faces for the purpose of identification and analysis are far from settled. IBM used 10 established "independent coding schemes from the scientific literature."[26] These schemes target craniofacial features and a number of subjective elements derived from human-labeled predictions for age and gender. While IBM's DiF dataset may be designed to help redress issues of diversity in training face recognition and expression analysis algorithms, the face coding schemes used to label its million faces still work with normalized coding schemes (developed by people) and subjective estimates. Ultimately, the apparent success of FaceApp, for instance, in artificially "aging" a face on the basis of schematic calculations remains just that – an estimate.

As the job of estimating identity, age, expression and other readable features of faces is handed over to seeing machines, enormous trust is placed in the underlying science of faciality. There are clear risks in this and, as we note later, significant pushback and even political resistance. However, from the beginnings of webcam and YouTube, Instagram and Snapchat, a perhaps irreversible complicity has helped to establish new norms around the way we use face images socially and as technical objects with a social functionality. This new mode of camera consciousness is underwritten by our complicity and intimate involvement in the dynamics of the face as a platform for communication and as a digital object of high value. And, so, in many ways, we are caught between rushing forward with the tech and recoiling in horror and anxiety at the speed at which we are becoming facial data within connected systems of smart cameras reaching out from social platforms to other contexts and spheres of life.

Platforming Facial Rekognition: Amazon and the Commerce of Faciality

There are many familiar tropes and tenets to the public and even political pushback against some uses of face recognition technology. And this pushback has gained momentum in the last two years. In many areas of the United States, political reluctance to wholly embrace face recognition for policing and governance has followed relentless pressure from a wide range of civil rights advocates and organizations. Interestingly, this has been increasingly backed in those areas by both major sides of politics. Concerns about the overreach, bias and inaccuracy of potentially discriminatory systems has resulted in a number of city-wide bans.

San Francisco was the first US city to ban use of face recognition technology by police and other agencies,[27] followed by Somerville, Massachusetts, and Oakland, California.[28] Advocacy and activist group Fight for the Future provides a detailed OpenStreetMap overview of where face recognition technology has been used or banned, and some of the government–private partnerships enabling its spread.[29] In the United Kingdom, developers overseeing work at King's Cross Station in central London have had to defend their use of face recognition without Local Council knowledge. As a busy transport and business node, it is not surprising that these revelations would raise public concern, but in doing so it has only emphasized how extensive the commercial uses of face recognition systems are, and the general lack of both government oversight and public awareness of their operation.[30]

Transparency is one of the vexed facets of the new camera consciousness generally, and is heightened in the case of automated face recognition. Our awareness of being visible to others is usually bounded by relationships among bodies, proximity and lines of sight. As smart cameras slip into the background of an urban landscape, so does any assuredness of the status of our visibility once tied only to the awareness of others. We are no longer anonymous in a crowd, or alone in a venue or premises or in what we might think to be an empty space, wherever smart cameras may be operating. But we aren't always aware of this. How can transparency be achieved? How can reciprocity be established in this landscape? For those who put them in place, security and governance applications are often taken by default to be in the public interest (except for where systems are proved inaccurate or ineffective).

Commercial applications of face recognition technology are shaped by the interests of the big tech developers, such as Amazon among others. In wealthy countries, much of the development is led by commercial interests, including applications seeking to make security more cost-effective and less labor-intensive. But a growing set of commercial and retail applications are also emerging as a means of tracking consumer movement within commercial spaces and generating insights into the complexities of consumer behavior. In his study of "emotional AI," Andrew McStay notes the commercial interests in applications ranging from neuromarketing and ad-testing to tracking customers' reactions to goods and services, tracking people in commercial spaces and personalizing service venues or digital products and service.[31] Eye-level camera tracking is also used within stores to gauge emotional responses to products, connecting "lingering" or "dwell" time in a store with aggregated demographics, gender, age, ethnicity and so on. The difficulty with these applications is more than just one of bias; it's the reach of their capability, the capacity of these systems to encompass us completely within integrated information systems that track and define our every move.

While civil rights groups and other advocates focus heavily on the problems of bias and the breach of privacy, attention should also be placed on the commercial forces that are accelerating both the development and application of face recognition. This current growth phase in the development and application of face recognition is facilitated by a set of processes referred to as "platformization." In their book *The Platform Society*, Van Dijck, Poel and de Waal understand platformization as the convergence of a number of key elements in the historical context of the rise of a big-tech infrastructural ecosystem. First, there is the introduction of systems for extended data collection and data linking, deployed by Internet platforms through APIs. This allows Facebook, Instagram or an app like FaceApp to interface with other software or share data and content.[32] Second, they also emphasize "the transformation of an industry where connective platform operators and their underpinning logic intervene in societal arrangements."[33]

Platformization has a social, economic and sectoral context and impact. Central, powerful nodes in the technology ecosystem – the "ecosystem builders," Facebook, Google, Amazon, Microsoft and Apple – operate both as infrastructures and service platforms on which others build and develop applications. They lay the foundation for systems that offer users (whether individuals, commercial or nonprofit organizations or state actors) "convenience in exchange for control over their data."[34] What Van Dijck et al. refer to as infrastructural platforms "can obtain unprecedented power because they are uniquely able to connect and combine data streams and fuse information and intelligence."[35]

Amazon Rekognition, the machine vision component of Amazon Web Services (AWS), received particular scrutiny and criticism in late 2018 and 2019. Employees and investors spoke out against the uses to which Rekognition was being put by clients in law enforcement and security, and investors reportedly applied pressure.[36] Rekognition is Amazon's version of the platformization of a face recognition, tracking and analysis service. With Rekognition, Amazon aims to build a service for others to run image and face recognition analysis on their own video or photo datasets. It simplifies the complex mathematics of machine learning for clients. As its website explains:

> You just provide an image or video to the Rekognition API, and the service can identify the objects, people, text, scenes, and activities, as well as detect any inappropriate content.[37]

The real-time services include face recognition and facial analysis, celebrity recognition, unsafe content detection and recognition of text within images and "pathing." Amazon explains pathing as capturing and tracking the path of people in a video scene; while Amazon offers the example of tracking the path of athletes or sports players, this process has further applications, such as tracking individuals in stores or other venues. Amazon's case studies attempt to show a range of potential applications for the service, yet there is a heavy emphasis on security.

Partly as a response to the sequestration of face image datasets and face recognition and analysis algorithms as valuable commercial property, a whole area of research has emerged around what is called "algorithmic auditing." This research tests the outputs of face recognition and other algorithmic processing to understand their outcomes for individuals. Mostly this work tests the performance of algorithms against their developer's base claims. So, for example, if a system claims to achieve ≥90% identification accuracy, auditing research will apply an audit framework to establish benchmarks and performance metrics against which that accuracy can be tested and biases can be identified.

Some of the backlash against Amazon's Rekognition platform came as a response to MIT research revealing unacceptable gender and ethnicity bias. In their published research, Raji and Buolamwini developed an audit design framework that presents performance metrics for IBM, Microsoft and Megvii (Face ++) face recognition systems in comparison with Amazon (Rekognition) and Kairos.[38] They found that "the overall performance of non-targets Amazon and Kairos lags significantly behind" that of the targeted companies IBM, Microsoft and Megvii when it came to recognition of gender and skin color, with error rates of 31.37% and 22.50%, respectively, for the darker skin female subgroup. The researchers also make strong arguments about the "real world impact" of making the results of such audits public. That is, they sought to find out whether the target platforms fix their biases and release new API versions.

There are key first-order ethical questions that arise from such blatant biases and error rates when such platforms are then used to target general populations for jobs like security and policing. Beyond the issue of error, there are more complicated second-order ethical questions that arise: how to build inclusivity into systems. But equally important is the question (probably yet to be answered) of what individual and social enhancements, augmentations and benefits face recognition can deliver.

The stakes are increasing rapidly as the face recognition platform arms race accelerates. Other big western technology "ecosystem builders," as Van Dijck and colleagues call them, have also established platform packages for AI services. IBM has Watson Visual Recognition. Google's Vision AI platform takes a broader focus than face recognition, showcasing services for automating drone vision industrial inspection, for example, but this platform also integrates with Google's extensive database of image and text, enabling powerful document classification and image search. Microsoft has released Windows Vision Skills, packages for AI-driven photo and video analysis tasks in April 2019. The modules include an object detector, skeletal detector and emotion recognizer components. These have been open sourced, as Microsoft developer Eliot Cowley explains, in order to "standardize the way computer vision models are put to use within a windows application."[39]

Facebook open sourced some of its machine vision tools around 2015–2016 through a BSD (Berkeley Source Distribution) license, which places few restrictions on reuse, on the software development platform GitHub. The Facebook AI Research Lab (FAIR) tools – DeepMask, SharpMask and MultiPathNet – work to segment objects within digital images by generating object masks, refining them and identifying the objects.[40] Facebook has long been able to accurately read faces in photo and video posts, automatically tag them accordingly and, at the same time, use that information to improve its social graph calculations about the relationships between people in a network. At the time of writing, Facebook is still fighting a lawsuit in the Illinois courts for breach of privacy under the state's Biometric Information Privacy Act, for collecting and using face data without clearly communicating its intent to its massive user base.

While these platform builders are open-sourcing machine vision tools and offering developer tools as a mechanism for helping to improve and inform those systems, the processes taken have so far done little to improve the transparency of the end-use or application. Historically, we are at the point where application and needs – whether commercial, state governance or public good related – may start to accelerate development through expanding economies of scale. But acceleration without transparency and reciprocity of value is precisely where public awareness and concern flare up.

The Face of the New Digital Citizen: China's Social Credit Program

As one of the most prominent and pervasive applications of face recognition for the purpose of mass automated governance, one thing is certain about China's social credit system: it puts into the spotlight the vexed role of vision and visibility in the organization of society and community. Outside of the most populous country in the world, China's social credit system stands in for not just a technical system of surveillance and governance but also for a regime that represents an imagined opposition to democratic freedoms. The construction of a pervasive surveillance camera network signals to many the extension of one particular dimension of the bio-mechanical functioning of vision and perception as an automated control system. However, within China, the experience of the multiple pilot projects and social governance models is not so straightforward or uniform.

A full account of the historical trajectory, and the localization, of these techno-social developments is vital to estimating the social transformation involved, but it is beyond the scope of this book. Essentially, the social credit system expands the objectives of localized monitoring and social and financial management through technological systems. The consequences for small infractions can appear shocking to those outside of China, and a likeness to Orwell's

Big Brother is an indicator of the association with brutal technology-assisted authoritarian rule. Samantha Hoffman, a researcher with the Australian Strategic Policy Institute, explains that, "for the system to function, it must provide punishments for acting outside set behavioural boundaries and benefits to incentivise people and entities to voluntarily conform, or at least make participation the only rational choice."[41] Citizens can be blocked from purchasing train or plane tickets, for instance, as a result of unpaid fines or other infractions.

The fast pace of urbanization in China has forced large-scale innovation in social organization and governance. Digitizing and automating aspects of this governance is by many accounts accelerating an autocratic and unreasonably invasive form of digital citizenship. If surveillance technologies are inherently geared toward forms of "social sorting," as Lyon puts it (2005), then China's social credit agenda can only be understood as conforming to the automation of a data-driven social control system. There have certainly been reports that highlight the move toward repressive technological governance. In a *New York Times* article, Paul Mozur details the use of face recognition for monitoring ethnic minority and mostly Muslim Uyghurs in its western regions.[42] This is the first known use of this technology for mass state-based racial profiling. While face recognition systems elsewhere in the world are continually criticized for embedding racial biases, racial difference is the primary target of China's use of the technology for tracking Uyghur people. The classification process in this case targets physical features such as skin tone and facial characteristics. Because Uyghur people are closer to central Asian in appearance than the majority Han population, authorities in the western regions have trialed the technology to track Uyghur people, though deny doing this for discriminatory or persecutory purposes. What is significant here is the capability intent, the willingness to train machine vision systems to pull individuals from a crowd on the basis of their racial profile.

Announced in 2014, the idea of a more centralized social credit system (historically operating as a number of different and often independent local governance systems) had been in development years before that, and with a longer trajectory in Communist Party led systems for management and control of social and business activity. The plan has been to realize a national reputation system through the collection and integration of extensive personal data, with machine learning and other data science techniques aiding the automation of the assessment of Chinese citizens' behavior and businesses' financial processes as an index of reputation. Like the integrated credit ratings in most Western countries, or social rating systems for eBay, Uber and similar systems of service and exchange, infringements will lower a citizen or business's score, while good deeds, acts of charity or voluntary work will help to raise it (Figure 3.1). However, it is really a larger set of financial, political and social control and management systems and technologies – many elements of which are still in a planning and development phase.[43]

FIGURE 3.1 Graphic illustration of face recognition technology and social credit score.

Source: Illustration by Kevin Hong.

Paper records of citizens and households had been used for similar purposes at a local government level for some time. The idea of the social credit system has its foundations in the China "Grid" social management system or "China National Grid" project. Under this framework, community watchers are employed within a social unit or segment within a local government area. In many countries around the world, this might align with a "neighborhood watch" program, which would serve a similar, albeit more limited, purpose of monitoring law, order and safety in a local area. In the grid management system, community watchers are assigned to report on "population size in the area, housing and facilities, social organizations, and other details," and identify any unusual activity.[44]

The more information or data that is available, the better an AI system works. In 2017, the BBC reported on China's ambitions to build the world's biggest camera-based surveillance system, with 170 million cameras in use, and plans to enlist 400 million more by 2020.[45] This is the novel element that provokes anxiety for many, fueled by China's meteoric rise in developing AI and machine vision capabilities. The system uses big data collection, integration and analysis alongside face recognition technology. And it is not without significance that the most sensational stories of its application have been the image and idea of public censure, where the faces and names of jaywalkers are projected on big screens at major urban intersections as both warning and social shaming technique. This is where the system's ambitions toward social management are most visible. Entrepreneurs, companies and individuals are shown the potential for an automated form of social and cultural accountability.

To what extent can automated camera technologies replace the eyes and ears of the hundreds of thousands of community watchers across China? Zhou Wang, an academic at Nankai University, writes about being quickly sized up by three such red armband-wearing women on his arrival to a northern Chinese city.[46] The job of inspecting and reporting on unusual activity or the presence of strangers in the city has fallen with these "community volunteers" under the grid management system. A small stipend and a reward for information that proves valuable keeps them engaged and on the lookout. Wang associates the implementation of grid management with the mass urbanization that has seen cities across China swell, creating what he calls a "society of strangers." As urbanization threatens social harmony, there arises a need to test and implement new surveillance techniques.

> Although Chinese cities are deploying a wide range of surveillance and facial identification technologies, at present most of these devices are only able to identify known threats to public order. In short, current technology is reactive, not preventative: A computer cannot judge the potential risk posed by an unknown individual, let alone determine what constitutes suspicious behavior.[47]

China's social credit system highlights the question of what it means to be a social actor, and to be visible to others and to the systems that have come to determine how the social imposes itself on the individual actor within. These processes are the often-contested basis for every form of citizenship. As Engin Isin and Evelyn Ruppert argue: "we cannot simply assume that being a citizen online already means something (whether it is the ability to participate or the ability to stay safe) and then look for those whose conduct conforms to this meaning."[48] We become digital citizens through every digitally mediated interaction. When this happens without awareness and reciprocity of oversight, both security and trust are compromised. This is why China's applications of face recognition for social governance raise the specter of wholesale social transformation.

Ubiquitous digital imaging[49] and machine vision disrupt the dynamics of seeing and being seen. This has deep implications for the social. Again, what happens when we are no longer anonymous in a crowd? Since at least the 1903 writings of Georg Simmel on the "Metropolis and Mental Life," the relationship between the individual and the social within urban contexts has brought both into question. Simmel's account of the move from rural settings to modern metropolis was one of the loss of identity, of a new kind of alienated, productive individuality characterizing the urban dweller's relationship with the modern machine. When trained toward individuals, machine vision systems are the latest stage of the reversal of this process, establishing exact points of identification and fine-grained segmentation. So much of what we understand as the social has been based on the pairing of human vision and conscious or deliberate self-determination. Brighenti's notion of the basis of society in seeing and being seen, while perhaps romantic, describes an understanding of the social as based in the visual confirmation of the self in relation to others.

As our relationship with networked, automated cameras changes, so does something of the edges of society (both the scope of the social, and the "links" that it consists of, in the network science sense). As the cameras themselves disappear into their everyday environments, use of face recognition technology builds a new camera consciousness that automates certain behaviors at a mass scale. As Josephine Wolff notes: "China is not just building up its database of people's faces, it is also embedding in people's minds the idea that face recognition is an essential and desirable part of everyday life."[50] There is an efficiency and accuracy trade-off between human-led and machine-led systems. While automated systems operate beyond the capacity of a human workforce, they still work within the parameters set by human and policy design. As a "banal" set of systems, smart, networked cameras are ubiquitously embedded in everyday contexts to help manage everything from the flow of people through airports and subways, the spread of viruses among livestock, attendance at university, to unlocking a phone or laptop, or making fast payments in stores. Where these become tools for social management and governance under the banner of "security," we do not yet know the effects on mental health or social interactions.

During the 2019 pro-democracy and anti-extradition law protests in Hong Kong, the fear of networked smart lampposts carrying face recognition technology was clear. The threat posed by those systems was palpable. With more than 2000 arrests, and the targeting of prominent protest leaders, the power over citizens posed by a distributed network of automated face recognition cameras played out a worst-case scenario. Frontline protesters wore black masks, eye protection and gasmasks, not only to protect themselves from physical harm, but also from view. The frontline protesters' tactic of choice was to disrupt the city and "disappear" quickly through the subway system. Scenes of umbrella holding groups cutting down smart lampposts circulated as a symbol of a tactical attack on the governing system that held a technological advantage. The Hong Kong company found by protesters to be associated with the lampposts, TickTack Technology, made a public statement that they would withdraw from the government contract.[51] While the use of face recognition through these street-side surveillance devices was denied, the destructive reaction of protesters signaled the awareness – the camera consciousness – that both subjugates and erupts in angry response.

Conclusion

Having traced a number of existing applications of and anxieties around face recognition, we close this discussion of face recognition by inquiring after possible reorientations of this technology toward achieving social good. Is there potential, for instance, in reversing the ledger, so to speak, by shifting the focus from the use of face recognition for governance and social control to addressing other issues affecting the social lives of citizens? Can the application of face recognition actually help to identify vulnerability, isolate it, and understand it more intelligently? That is to say, can this technology be reoriented around a social change agenda, so as to target resources and services toward those most in need in society. Can it be trained to automate forms of redress for social suffering, rather than simply to effect and expedite policing strategies? Can it help to identify the societal (rather than profile oriented) drivers of crime and add to crime prevention strategies? Can the same mechanisms that learn to identify, individuate and isolate people within a crowd be used to understand and help to tackle social exclusion and build social inclusion? Our modest proposal here is that, while the potential of face recognition technologies and systems to create new social exclusions has been and is enormous, and is being realized in China (and at any border or airport using face recognition), it might also be feasible to see in China's agenda for face recognition an important and as yet little explored corollary or alternative: use of face recognition as a form of computational responsiveness to individuals, which could incorporate explicit social inclusion interventions and considerations.

As we have seen in this chapter, the value of visual data expands exponentially when it becomes attached to, or comes to define key aspects of, bio-demographic information, which is why face recognition systems might form a key part of an expanding social credit system. Given these expanding contexts of use, it is valuable to remember and further consider the double element of the camera and its socializing abilities. On the one hand, there are anxieties or power relations the camera creates in its capacity to see and set relationships before it. And, on the other hand, there are camera's capacities for navigation, discovery, witnessing and its ability to extend perception and affect sociality. Image-making is also wielded as an outcome of the mobile, smart camera moment in a way that has dramatically reshaped the social and the individual's place and agency within it. This reshaping process that follows the advent of smart cameras forms the focus of the following three chapters, beginning with an examination of the extension of vision and visibility made possible by mobile phones.

Notes

1. Gilles Deleuze and Felix Guattari, *A Thousand Plateaus: Capitalism and Schizophrenia*, trans. Brian Massumi (London: Continuum, 2002), 190.
2. Paul Ekman and Wallace V. Friesen, *Unmasking the Face: A Guide to Recognizing Emotions from Facial Clues* (San Jose, CA: ISHK, 2003); Grant Bollmer, "Books of Faces: Cultural Techniques of Basic Emotions," *NECSUS*, Spring (2019), https://necsus-ejms.org/books-of-faces-cultural-techniques-of-basic-emotions/; Andrew McStay, *Emotional AI: The Rise of Empathic Media* (London: Sage, 2018).
3. Kelly A. Gates, *Our Biometric Future: Face Recognition Technology and the Culture of Surveillance* (New York: NYU Press, 2011), 23–24.
4. Ibid., 24, 26, 29.
5. Deleuze and Guattari, *A Thousand Plateaus*, 195.
6. Roland Barthes, *Mythologies*, trans. Annette Lavers (New York: Hill and Wang, 1972).
7. Josephine Wolff, "China's Push Towards Face Recognition Technology," *China – US Focus*, March 12, 2019, www.chinausfocus.com/finance-economy/chinas-push-towards-facial-recognition-technology.
8. José van Dijck, Thomas Poell, and Martijn de Waal, *The Platform Society: Public Values in a Connective World* (London: Oxford University Press, 2018).
9. Nathan Jurgenson, *The Social Photo: On Photography and Social Media* (New York: Verso, 2019); Theresa M. Senft, *Camgirls: Celebrity and Community in the Age of Social Networks* (New York: Peter Lang, 2008).
10. Katharina Lobinger, "Photographs as Things – Photographs of Things. A Texto-material Perspective on Photo-sharing Practices," *Information, Communication & Society* 19, no. 4 (2016): 475–488; Theresa Senft and Nancy K. Baym, "Selfies Introduction – What Does the Selfie Say? Investigating a Global Phenomenon," *International Journal of Communication* 9 (2015): 1588–1606, https://ijoc.org/index.php/ijoc/article/view/4067/1387; Katrin Tiidenberg and Edgar Gómez Cruz, "Selfies, Image and the Re-making of the Body," *Body & Society* 21, no. 4 (2015): 77–102; Amparo Lasén, "Digital Self-Portraits, Exposure and the Modulation of

Intimacy," in *Mobile and Digital Communication: Approaches to Public and Private*, J.R Carvelheiro and A.S Telleria (Eds.) (Covilhã, Portugal: Livros LabCom, 2015).

11. Ibid., 65.

12. Aaron Hess, "Selfies| the Selfie Assemblage," *International Journal of Communication* 9 (2015): 1631, https://ijoc.org/index.php/ijoc/article/view/3147/1389.

13. Stuart Dredge, "Five of the Best Face Swap Apps," *The Guardian*, March 17, 2016, www.theguardian.com/technology/2016/mar/17/five-of-the-best-face-swap-apps.

14. Paige Leskin, "Since Going Viral Again for Making People Look Old, FaceApp Has Been Downloaded by 12.7 Million New Users," *Business Insider*, July 19, 2019, www.businessinsider.com.au/faceapp-viral-downloads-13-million-new-users-last-week-2019-7?r=US&IR=T.

15. See, for example: Christie Wilcox, "Why FaceApp's Selfie Filters Work So Well, and Why They Don't," *Gizmodo*, May 10, 2017, www.gizmodo.com.au/2017/05/why-faceapps-selfie-filters-work-so-well-and-why-they-dont/.

16. Harold Stark, "Introducing FaceApp: The Year of the Weird Selfie," *Forbes*, April 25, 2017, www.forbes.com/sites/haroldstark/2017/04/25/introducing-faceapp-the-year-of-the-weird-selfies/#334634e243d2.

17. Google Play Store Review, 18/7/2019.

18. Hanbyul Choi, Jonghwa Park, and Yoonhyuk Jung, "The Role of Privacy Fatigue in Online Privacy Behavior," *Computers in Human Behavior* 81 (2018): 42–51.

19. Robert W. Gehl, "The Archive and the Processor: The Internal Logic of Web 2.0," *New Media & Society* 13, no. 8 (2011): 1228–1244.

20. See, for example: "Face Recognition Homepage," www.face-rec.org/databases/.

21. Kate Crawford and Trevor Paglen, "Excavating AI: The Politics of Images in Machine Learning Training Sets," *excavating.ai*, September 19, 2019, https://excavating.ai.

22. John R. Smith, with Joy Buolamwini and Timnit Gebru, "IBM Research Releases 'Diversity in Faces' Dataset to Advance Study of Fairness in Face recognition Systems," *IBM Blog*, January 29, 2019, www.ibm.com/blogs/research/2019/01/diversity-in-faces/.

23. Russell Brandom, "Microsoft Pulls Open Face Recognition Dataset after *Financial Times* Investigation," *The Verge*, June 7, 2019, www.theverge.com/2019/6/7/18656800/microsoft-facial-recognition-dataset-removed-privacy.

24. Olivia Solon, "Face Recognition's 'Dirty Little Secret': Millions of Online Photos Scraped Without Consent," *NBC News*, March 12, 2019, www.nbcnews.com/tech/internet/facial-recognition-s-dirty-little-secret-millions-online-photos-scraped-n981921.

25. Smith, "IBM Research Releases."

26. Leslie G. Farkas, *Anthropometry of the Head and Face* (New York: Raven Press, 1994); Alain Chardon, Isabelle Cretois, and Colette Hourseau, "Skin Colour Typology and Suntanning Pathways," *International Journal of Cosmetic Science* 13, no. 4 (1991): 191–208; Yanxi Liu, Karen L. Schmidt, Jeffrey F. Cohn, and Sinjini Mitra, "Facial Asymmetry Quantification for Expression Invariant Human Identification," *Computer Vision and Image Understanding* 91, no. 1–2 (2003): 138–159; Leslie G. Farkas, Marko J. Katic, and Christopher R. Forrest, "International Anthropometric Study of Facial Morphology in Various Ethnic Groups/Races," *Journal of Craniofacial Surgery* 16, no. 4 (2005): 615–646; Narayanan Ramanathan and Rama Chellappa, "Modeling Age Progression in Young Faces," *International Conference on Computer Vision and Pattern Recognition (CVPR)* (New York: IEEE, 2006), 387–394, https://doi.org/10.1109/CVPR.2006.187; Anthony C. Little, Benedict C. Jones, and Lisa M. DeBruine, "Facial Attractiveness: Evolutionary Based Research," *Philosophical Transactions of the Royal Society* 366 (2011): 1638–1659; Xiangxin Zhu and Deva Ramanan, "Face Detection, Pose Estimation, and Landmark Localization in the

Wild," *International Conference on Computer Vision and Pattern Recognition (CVPR)* (New York: IEEE, 2012), 2879–2886; Aurélie Porcheron, Emmanuelle Mauger, and Richard Russell, "Aspects of Facial Contrast Decrease with Age and Are Cues for Age Perception," *PLoS One* 8, no. 3 (2013), https://doi.org/10.1371/journal.pone.0057985; Ziwei Liu, Ping Luo, Xiaogang Wang, and Xiaoou Tang, "Deep Learning Face Attributes in the Wild," *International Conference on Computer Vision (ICCV)* (New York: IEEE, 2015), 3730–3738; Rasmus Rothe, Radu Timofte, and Luc Van Gool, "Deep Expectation of Real and Apparent Age from a Single Image Without Facial Landmarks", *International Journal of Computer Vision* 126, no. 2–4 (2018): 144–157.

27. Kate Conger, Richard Fausset, and Serge F. Kovaleski, "San Francisco Bans Face Recognition Technology," *New York Times*, May 14, 2019, www.nytimes.com/2019/05/14/us/facial-recognition-ban-san-francisco.html.

28. Shirin Ghaffary and Rani Molla, "Here's Where the US Government Is Using Face Recognition Technology to Surveil Americans," *Vox / Recode*, July 18, 2019, www.vox.com/recode/2019/7/18/20698307/facial-recognition-technology-us-government-fight-for-the-future.

29. Ibid.; "Fight for the Future" (2019), www.fightforthefuture.org/; "Ban Face Recognition Map" (n.d.), www.banfacialrecognition.com/map/.

30. Zoe Kleinman, "King's Cross Developer Defends Use of Face recognition," *BBC News*, August 12, 2019, www.bbc.com/news/technology-49320520; Martha Busby, "People at King's Cross Site Express Unease About Face Recognition," *The Guardian*, August 14, 2019, www.theguardian.com/technology/2019/aug/13/people-at-kings-cross-site-express-unease-about-facial-recognition.

31. McStay, *Emotional AI*.

32. Anne Helmond, "The Platformization of the Web: Making Web Data Platform Ready," *Social Media + Society*, July–December (2015): 1–11, https://doi.org/10.1177/2056305115603080.

33. Van Dijck, Poell, and de Waal, *The Platform Society*, 169, note 18; Jean-Christophe Plantin, Carl Lagoze, Paul N. Edwards, and Christian Sandvig, "Infrastructure Studies Meet Platform Studies in the Age of Google and Facebook," *New Media & Society* 20, no. 1 (2018): 293–310.

34. Van Dijck, Poell, and de Waal, *The Platform Society*, 16.

35. Ibid., 16.

36. Natasha Singer, "Amazon Faces Investor Pressure Over Face recognition," *New York Times*, May 20, 2019, www.nytimes.com/2019/05/20/technology/amazon-facial-recognition.html; Anna Merlan and Dhruv Mehrotra, "Amazon's Facial Analysis Program Is Building a Dystopic Future for Trans and Nonbinary People," *Jezebel*, June 27, 2019, https://jezebel.com/amazons-facial-analysis-program-is-building-a-dystopic-1835075450.

37. "Amazon Rekognition," *Amazon Web Services*, https://aws.amazon.com/rekognition/.

38. Inioluwa Deborah Raji and Joy Buolamwini, "Actionable Auditing: Investigating the Impact of Publicly Naming Biased Performance Results of Commercial AI Products," in *Proceedings of the 2019 AAAI/ACM Conference on AI, Ethics, and Society (AIES-19)* (New York: ACM, 2019), https://dam-prod.media.mit.edu/x/2019/01/24/AIES-19_paper_223.pdf.

39. Kyle Wiggers, "Microsoft Releases Windows Vision Skills Preview to Streamline Computer Vision Development," *VentureBeat*, April 30, 2019, https://venturebeat.com/2019/04/30/microsoft-releases-windows-vision-skills-preview-to-streamline-computer-vision-development/.

40. Pedro O. Pinheiro, Ronan Collobert, and Piotr Dollár, "Learning to Segment Object Candidates," in *Proceedings of the 28th International Conference on Neural Information Processing Systems – Volume 2* (New York: ACM, 2015), 1990–1998; Pedro O.

Pinheiro, Tsung-Yi Lin, Ronan Collobert, and Piotr Dollár, "Learning to Refine Object Segments," in *European Conference on Computer Vision* (Cham, Switzerland: Springer, 2016), 75–91); Sergey Zagoruyko, Adam Lerer, Tsung-Yi Lin, Pedro O. Pinheiro, Sam Gross, Soumith Chintala, and Piotr Dollár, "A Multipath Network for Object Detection," *arXiv preprint* (2016), https://arxiv.org/pdf/1604.02135.pdf.

41. Samantha Hoffman, "Social Credit," *Australian Strategic Policy Institute*, June 28, 2018, www.aspi.org.au/report/social-credit.

42. Paul Mozur, "One Month, 500,000 Face Scans: How China Is Using A.I. to Profile a Minority," *New York Times*, April 14, 2019, www.nytimes.com/2019/04/14/technology/china-surveillance-artificial-intelligence-racial-profiling.html.

43. Hoffman, "Social Credit."

44. Yongshun Cai, "Grid Management and Social Control in China," *Asia Dialogue*, April 27, 2018, https://theasiadialogue.com/2018/04/27/grid-management-and-social-control-in-china/.

45. Wolff, "China's Push."

46. Zhou Wang, "Is China's Grassroots Social Order Program Running Out of Money?" *Sixth Tone*, trans. Kilian O'Donnell, June 1, 2018, www.sixthtone.com/news/1002393/is-chinas-grassroots-social-order-project-running-out-of-money%3F.

47. Ibid.

48. Engin Isin and Evelyn Ruppert, *Being Digital Citizens* (London: Rowman and Littlefield International, 2015), 19.

49. Martin Hand, *Ubiquitous Photography* (Cambridge: Polity, 2012).

50. Wolff, "China's Push."

51. Holmes Chan, "Hong Kong Tech Firm Pulls Out of Smart Lamppost Programme After Surveillance Accusations and Staff Threats," *Hong Kong Free Press*, August 26, 2019, www.hongkongfp.com/2019/08/26/hong-kong-tech-firm-pulls-smart-lamppost-programme-surveillance-accusations-staff-threats/.

4

AUTOMATING AND AUGMENTING MOBILE VISION

Over a relatively short period, mobile phones have become powerful minicomputers, ubiquitous, full of complex sensors. If drones stand as one of the most visible and dramatic figures of automated machine vision, the mobile phone hides in plain sight as it datafies everyday life like never before. While many of the social transformations associated with mobile communication and mobile media are well documented,[1] our point here is that the successive technological innovations driving new iterations of mobile phones coalesce around a particular drive: they all work to extend and alter the terrain and functionality of social visibility, from the pocket or hand of anyone, wherever they are. As key actors in our new media ecosystem, mobiles continue to revise social relations and the spaces and times of everyday life. Perhaps mobile devices are no longer considered figures of "enchantment," but they remain "vital objects" as Jane Bennett puts it,[2] distributing mobile and social imaging[3] and generating dynamic maps on a global scale.

For such a small device, mobile phones can pack an incredible array of sensors, including accelerometers, gyroscope, magnetometer, proximity sensor (e.g., so that the screen goes off when it's up to ear or turned down on a desk), light sensor, GPS and barometer (to improve GPS accuracy), air humidity sensor, heart rate monitor, finger print scanner, eye and facial scanners, Wi-Fi and Bluetooth locators and so on. And, visual processing power associated with the camera multiplies the mobile phone's seeing and sensing capacity even further. In fact, mobile cameras and image processing hold the key to the next wave of augmented (AR) and virtual reality (VR) applications. Simple object awareness systems are already searching for social applications.

One example of a mobile phone's object and spatial awareness capabilities, Apple's iBeacon, works with location services to allow a phone's operating system to connect location-based apps with objects or places and, for example, trigger alerts when people approach. It doesn't use latitude and longitude to determine location, but rather makes connections through a low-energy Bluetooth signal. This brings into play a simple mode of indoor positioning with applications in assistive technologies for vision-impaired people, in-store automated alerts, media pushing or transactions, for example. These capabilities are likely to undergo even further refinement by Apple with the introduction of its UWB (ultra-wideband) U1 chip, which will be able to pinpoint the location of a device to within a few centimeters.[4] Meanwhile, Microsoft and Google are racing to render the material world searchable through cloud-based visual processing and machine learning.[5] The possible applications for each of these uses of mobile phones for geospatial communication are vast, if yet to be fully realized, and the human implications as yet not well understood.

Well embedded in everyday life, mobile phones continue to reconfigure the way we navigate urban environments, access information, act socially, connect with others and share or experience everyday moments.[6] They layer our urban environments as navigable information ecosystems.[7] But, like drones and driverless cars, they also pose questions of what it means for a machine or device to sense and process, or "understand" and "remember," its environment, and how our personal devices extend the field of visibility. How does a mobile phone act socially and operate as a camera conscious seeing machine? How does it extend seeing into sensing, and how does that sensing extend the terrain of visibility and introduce new functionality as material spaces become geospatial information systems? Moving away from the predominant focus on camera phone practices and ambient imaging, our approach is to rethink mobile phones as they affect a new camera consciousness.

Mobile imaging and visual processing technologies play an interesting role in reshaping everyday visibilities. We can see this in developments in internal mapping as it is imagined and tested through Google's Project Tango mobile devices. This project is emblematic of the trajectory of personal mobile media involving sophisticated sensing and visual processing power. Mobile devices are quickly becoming platforms for expanded visual-based AR and VR systems, and open new opportunities for developers looking to recast and repurpose the visible terrain of the social. As technologies of the new camera consciousness, mobiles have for some time now exemplified the expanded capacity to *see* through distributed points of view; added to this capacity are intelligent cameras and associated sensors that are networked and coupled with machine vision and cloud computing. In this sense, mobiles provoke questions about *how* individual local environments become visible and socially available; they also extend the possibilities of seeing and acting digitally beyond human senses. In this chapter, then, we open by tracing a number of key historical developments in camera automation. We then shift the emphasis from a focus on the histories and acts of photography and visual communication to examine the relative autonomy of mobile imaging, and image processing, which indicate the emerging value of visual data.

Historical Developments in Camera Automation

Contemporary automated mobile vision capture technologies have been a long time in the making. In order to fully comprehend and appreciate the impact and importance of modern-day smartphones as powerful minicomputers and as automated vision capture machines, it is valuable to look back at the history of photography and camera development. A historical perspective is productive here given that, as Daniel Palmer argues, "the history of photography is also a history of automation."[8] In the sections that follow, we explore the gradual introduction and incorporation of increasing autonomy within photography, which we trace across what Chris Chesher has termed the "dominant mass imaging traditions": film cameras (1899–2000), digital cameras (1986–), camera phones, feature phones with cameras (1992–) and smartphones (2007–).[9] Our historical account of these technological developments is intended to be diachronic rather than synchronic; that is to say, our aim is not to provide an exhaustive history of these developments but to note a number of key moments within a longer history of photographic automation.

Film Cameras

One key driver of automation has been to reduce human labor time.[10] And, as business efficiency experts have noted, if an organization simplifies a process prior to automation, the greater rewards are likely to be from the automation process.[11] Eastman Kodak was one of the first photographic companies to grasp the importance of this and of simplifying – and effectively automating – complicated photographic processes for end consumers. Eastman Kodak recognized that "the biggest profits would come from driving consumers to take more pictures, and the key to that was simplicity and ease of use."[12] To this end, in 1888, the Eastman Kodak Company released the Kodak, a box camera that retailed for US$25 and which came preloaded with a 100-exposure roll of film. As Reese Jenkins explains,

> The novice photographer had only to point the camera toward the desired subject and "push the button." When he had exposed the film, he had only to return the camera to the factory where for $10 the film was removed and replaced with a fresh roll of film and the exposed film processed.[13]

Supported by the famous marketing slogan, "You push the button, we do the rest,"[14] Eastman Kodak's aim with the Kodak camera was to alleviate the anxieties of amateur photographers "who feared photography's technical complexity."[15] In October 1901, the company produced a version of the Kodak camera pitched at children – the famous box Brownie; this was an affordably priced (US$1) camera that also came with preloaded film.[16]

By 1938, Eastman Kodak released the Super Kodak Six-20, which has been described as "the world's first automatic exposure camera."[17] The Super Kodak Six-20 used "a large selenium photocell to control the aperture before the exposure was made."[18] And, in the early 1960s, the company introduced the Kodak Instamatic 100, a "simple little plastic box camera with fixed focus, a pop-up flash gun, and rapid lever wind";[19] in addition to these features, its most significant attribute was a drop-in film cartridge, a popular innovation that is said to have "revolutionized the snapshot industry."[20]

Kodak, of course, was by no means the only firm working to simplify – and to automate – photographic processes. In 1947, the Massachusetts-based Polaroid Corporation demonstrated its Land Camera Model 95, billed as the "world's first instant camera."[21] While later versions of Polaroid cameras (which continue to have novelty appeal today) permitted prints to be developed outside the camera, the first version, the Land Camera Model 95, used roll film and produced finished prints in-camera within a minute:

> After exposure, pulling a tab caused the roll of negative film to join with print paper as they were drawn through rollers. At the same time the rollers spread developer evenly across the interface surface to process the print.[22]

A major step toward the development of electronic camera automation came in 1964 with the release, by the Tokyo-based Asahi Optical Company, of the Pentax Spotmatic camera (officially known as the Asahi Pentax SP). This was the first SLR (single-lens reflex) camera to offer "what would soon become standard on all SLRs: through-the-lens metering,"[23] whereby the intensity of light reflected from the scene being photographed is measured through the lens rather than through a separate metering window or an external handheld light meter. And, in 1971, Nikon revealed its High Speed Motor Drive, a unit that attached to the bottom of the Nikon F camera and which was capable of advancing roll film at seven frames per second.[24]

From the mid-to-late 1970s, developments in electronic camera automation began to multiply and accelerate. In 1976, Canon began production of the AE-1, the first 35-mm SLR camera to use a central processing unit (CPU); the CPU "regulated operations like exposure memory (known then as exposure lock) as well as aperture value control and the self-timer."[25] The following year, camera maker Konica released its C35AF model, "the world's first production autofocus camera."[26] In 1978, Polaroid announced the SX-70 Sonar OneStep, the first point-and-shoot autofocus camera,[27] while Leica demonstrated "the first SLR with fully operational autofocus."[28] Several years later, in 1981, the

Pentax ME–F was launched, forming the first commercially available autofocus 35-mm SLR camera.[29]

Digital Cameras

Building on the aforementioned developments, fully digital cameras began to emerge from the mid-1980s. In 1986, for example, Kodak "introduced a 1.4 megapixel sensor,"[30] while, in 1988, photographic film maker Fujifilm launched "the world's first fully digital consumer camera," the FUJIX DS-1P.[31] Digital camera innovation accelerated throughout the late-1980s and 1990s and continues into the present.[32]

John Berger once famously remarked that "the camera relieves us of the burden of memory."[33] As we have seen from the aforementioned abbreviated account of key moments in the history of camera automation, camera companies and photographic film producers have also sought to relieve us of the burden of other forms of photographic decision-making by automating many of these functions. According to Daniel Palmer, "the automation of factors like focus and exposure have typically been sold on the basis that they enable photographers to concentrate on responding instantly to the world around them, rather than investing unnecessary energy engaged in mastering unwieldy technology."[34]

It is also important to note that, just as the aforementioned innovations in camera technologies contributed to increased automation of cameras, parallel developments have also led to dramatic shifts in how we think about the photographic image itself. Photographs, Estelle Blaschke observes, are now seen not only as "an image carrier but as a carrier of data."[35] Blaschke makes the point that, as with camera automation, the cherishing of images as carriers of information has a long and rich history, which she traces back to the early 1800s.[36] What has shifted with the rise of digital cameras and "with coding and the use of electronic data management [tools, is that] the informational value of an image has achieved a new quality"[37] and a high level of industry and end user acceptance – so much so that metadata have come to be seen as "intrinsic to digital photography, and are embedded in every JPEG file"[38] and other compression codecs. In this way, digital photography, Blaschke argues, "has created image infrastructures that are [now] vital for the very existence of an image in the digital ecosystem."[39] A key outworking of this is that today the data that accompanies an image, or is embedded in and can be extracted from it, have become just as, if not more, valuable than the image itself.[40]

Camera Phones

The first camera phone – that is, a mobile feature phone with an embedded camera – was the Sharp J-SH04. This was released in Japan in 2000, and included a "rudimentary 0.1 megapixel camera."[41] Late the following year, Nokia released its 7650 model, the first GSM (Global System for Mobile Communication) mobile phone with a camera.[42] Within five years, almost all mobile cellular phones came with small, cheap CMOS (complementary metal-oxide semiconductor) sensor cameras.[43] For the most part, the image quality of these camera phones "was almost universally poor," due to "limitations in lens-depth, sensor size, limited storage and low cost electronics."[44]

While image quality might have been low, camera phones nevertheless came to be regarded as significant for a range of reasons. First, they further democratized picture taking, so that, "instead of many members of a family sharing one camera, they [could] now all photograph with their individual mobile phones."[45] Second, camera phones (or stand-alone digital cameras with wired or wireless connections) "exist as devices within a communications network,"[46] a point that has important implications for the wider circulation of digital images. And, third, their compact size and ease of use led to an explosion in popularity of camera phones and their incorporation into daily mobilities and routines.

Since the near ubiquitous uptake of mobile phones (and later smartphones) with camera equipment, a wealth of research has explored their incorporation into and impact on everyday life. A predominant focus on camera phone *practices* has emerged, represented by Kato, Okabe, Ito and Uemoto's seminal study of the "uses and possibilities of the Keitai camera" in Japan.[47] When the digital image becomes a fundamental unit of communication, of "perpetual visual contact,"[48] its value and potential as dynamic visual data multiplies. Visuality becomes "emplaced,"[49] for instance, and a mode of everyday "chit-chat,"[50] while redefining distinctions between public and private.[51] Lobinger[52] details the shifting grounds of "networked photography" and "social cameras"[53] that contribute to the distributed and connective nature of the contemporary digital image. With the later arrival of smartphones, these studies have been extended by accounting for the autonomous production of visual information and metadata, especially geo-locative data, that operationalize camera phones in new ways.

What this literature points to is the multiple ways mobiles are caught up in personal and social relations that affect their technical development, and which they in turn affect. It also points to how these camera-equipped phones have contributed to shifting expectations around image-taking and to the rise of what Susan Murray terms an "everyday aesthetic,"[54] where the primary aim of camera phone photography appears to be to "arrest the ephemerality of daily life, however fleetingly."[55] Everyone, it would seem, could now act as a modern-day Gregory Bateson, with a camera about their person, ready and prepared to photograph whatever everyday object or fleeting ritual that captured their anthropological fancy. There are also phenomenological considerations that come into play in the augmentation and anxiety surrounding personal mobile cameras, and the "material experience of vision" that Heidi Rae Cooley argues "results as hands, eyes, screen and surroundings interact and blend in syncopated fashion."[56] Phenomenological approaches remain vital to this field of study,[57] and have also been explored in relation to the development of a "film consciousness."[58]

Fundamental to camera phone developments is the associated goal of "visiblizing," or rendering the otherwise invisible visible across many spheres of social life. The Italian sociologist Andrea Brighenti offers insights into these processes and recounts the rich theoretical perspectives probing the conditions of contemporary visibility – as a theory of the social.[59] He examines visibility as oscillating between "recognition" and "control," pointing to a deep ambivalence in the context of mobile social media where practices of seeing and being seen become significantly reconfigured – rewarded but also anxiety provoking.[60] Brighenti takes this tension further than most to describe visibility as all the perceptual events, acts, relations that constitute a territory of the social.[61] He calls his approach an ecological phenomenology – where the visible operates as an "'open field' or an 'element' in which the social occurs";[62] that is, it is not an on/off quality, or unidirectional, and is produced socially as well as technically. This is a relational and pragmatic approach to visibility and visual perception, where visibility is "a property of the social field"[63] formulated in relation to the interactions of perceiver and perceived (whether human or non-human) and through techniques of mediation. As with Deleuze's notion of the semi-subjectivity of the camera, the concept of visibility formulated here bypasses the dichotomies that are assumed by many approaches to photography and camera phone practices that place more emphasis on the human rather than machine image maker.

It's difficult to think about the mobile camera beyond the practices of its user, although the technological developments we explore later illustrate why it's vital to do so. Mobile devices, through wireless networks, reshape the territory of the visible – they visiblize – in ways that are increasingly imperceptible. Of course, they multiply points of view through camera phone practices, but they are also saturated with automated sensors that visiblize in different ways. For example, a mobile phone connected to a wireless network via phone tower transmitters "notices" and is "noticed by" each tower as it moves across physical landscapes and over the fuzzy borders that the towers create *in relation to* that mobile phone. The "feeling of incompletion or openness" that Adrian Mackenzie describes of wireless networks[64] also underpins the capacity of a mobile device to affect the territory of the visible. Our approach highlights the relational capacity of the visible world and modes, techniques or technologies of visualization, which are especially relevant for the wireless technologies underpinning mobile phones. To this we add an account of visual data and digital image processing as core components of the new camera consciousness in which mobiles can be located.

Smartphones

The arrival of the smartphone – especially the iPhone – marked a further and very significant step in the path of development of mobile vision capture technologies. While the story of the iPhone is by now well known,[65] certain aspects of this story bear repeating in order to grasp the impacts of the iPhone, and smartphones in general, on mobile image-making. Chris Chesher notes that "when the first model of the iPhone was released in 2007, the camera seemed unlikely to change the world. It was relatively low resolution (2 megapixels), fixed focus and the default app couldn't even zoom the image."[66] And, yet, the initial appeal of this device was immense, with over 100 million iPhones sold in the first four years of its release.

The early success of the iPhone, and its framing as a "disruptive" imaging technology, can be attributed to a range of factors, including (but not limited to) its *hardware*; how it (and its camera) *operates within an app ecosystem*; how it *serves as a communication and a multi-media, data-driven device*; its revolutionary *touchscreen interface*; the fact that it has *GPS capability*; and how the camera serves as just one among a *range of "sensing" technologies*. Each of these elements will be touched on in turn.

As with camera phones, the iPhone deployed CMOS image sensors. What set the iPhone apart from existing camera phones, though, was its powerful processor. With the iPhone, photographic scenes could be "quickly grabbed – the fast processor gives the iPhone less shutter lag than most camera phones – and just as quickly and efficiently deleted, shared and archived."[67]

Also significant, as Daniel Palmer points out, was the fact that the native iPhone camera app was "just one piece of software among hundreds that can engage the camera hardware on the phone."[68] The situation of the iPhone within an app ecosystem meant that it was possible to access the camera through a wider array of photo apps beyond the native camera app. Palmer suggests that, broadly speaking, most photo apps (at least in the early days of the iPhone and app store) served two main purposes: making images "more 'artistic' or aesthetically appealing" and/or facilitating the wider distribution of images.[69] With respect to the first function, what these apps entailed was "a form of 'processing' of the JPEG image."[70] For example, in the case of Hipstamatic, a popular early iPhone photo app, this processing involved emulation of a retro look, which typically involved "a square format, faded tones, vignetting and chromatic aberration effects."[71] One suggestion is that image processing of this sort has contributed to "a minor re-invention of the amateur camera,"[72] and the formulation of a particular aesthetic form or mode of picture-taking, known as "iPhoneography."[73] With respect to the second function, much of the traffic in digital images now originates from smartphones and is disseminated and circulates through digital communication networks and is supported by image-sharing platforms.[74] Also, while the iPhone entered into an ecology where image-making had already transitioned from being print-focused to being screen- and transmission-focused, thanks to the prior arrival of digital cameras and camera phones,[75] it is worth noting that the iPhone's striking back-light display and touchscreen interface − now a feature of all smartphones − played an important role in redefining how photos were taken, displayed and shared with others.[76]

Further to the aforementioned, the GPS capabilities of the iPhone have meant that photographs can be tagged automatically with location, which allows images to be "browsed and arranged geographically."[77] Geodata has added an additional, important layer to the existing forms of metadata already captured within digital photographs. The significance of this is captured in what Chris Chesher refers to as the smartphone camera's "deictic" function, whereby information about a person, location, time and so forth are associated with and imbedded in images as part of their metadata.[78] In short, digital photographs have become "hybrid forms" that merge the "image and data *about* that image."[79] It is in this sense that digital images have shifted from being primarily about *representation* to increasingly about *information*, and where the camera is just one among many of the smartphone's sensors.[80]

An understanding of the smartphone and its camera as sensor carries pro-found impacts. First, it complicates distinctions between smartphones and the sorts of pilotless drones discussed earlier in this book. In Mark Andrejevic's account of drone theory, the mobile phone plays a central role in embedding a "drone logic" into everyday life.[81] Both the drone and the mobile phone can be understood in terms of their overlapping sensor arrays. They extend the reach of the senses and add new capacities. They saturate sensory time and space, automate sensemaking (through increasingly cloud-based data storage, processing and analytics) and automate responses − from adjustments to screen brightness, to automated object avoidance maneuvers, alarms, alerts, payments and other functions of geospatial awareness. For Andrejevic, the connection between drones and mobiles isn't accidental, or simply metaphorical, it is also socio-political, entailing a kind of "persistent surveillance" unique to our contemporary information society and the informational underpinnings of contemporary capitalism. Mobile phones have begun to act socially, not just for the purposes of personal communication and entertainment. Through image processing, sensor and activity mapping, they are continuously productive of visibility as the digital territory of the social.

Second, an understanding of the smartphone and its camera as sensor unsettles long-established understandings of photography as fundamentally a representational medium.[82] It also unsettles, or de-centers, the role of the photographer, insofar as digital images are increasingly aggregated, analyzed and acted upon by computers rather than just humans:

> A digital camera is not simply a passive recording device. It doesn't *take* pictures; it *makes* them. When the sensor intercepts a pattern of illumination, that's only the start of the process that creates an image.[83]

More-and-more this process is occurring independent of human involvement[84]; increasingly, cameras are "machine-to-machine seeing apparatuses."[85] In short, "computers are now cameras,"[86] and "cameras are now computers"[87] – "computers with sensors."[88] With the creation of the smartphone, Daniel Palmer writes, we are witnessing "a new wave in computational photography, in which cameras are increasingly able to recognize and interpret scenes and the resulting images are used to 'do' things."[89]

It is these capabilities – the abilities of cameras to recognize and interpret scenes and act upon them – that we explore in the remaining sections of this chapter. These aspects of the transformation of the territory of the visible can be understood through the examples of developments in visual processing chips for smartphones, of particular projects and applications that utilize these chips around indoor mapping and, finally, ambitious start-ups that attempt to provide an analytical "crystal ball" through the aggregation and processing of masses of mobile, located, metadata-rich social images.

Visual Processing Chips

Sometimes the biggest innovations are the smallest. Take visual processing chips, for instance. The kind of high definition, real-time image processing used by a drone to relay first person view images to a mobile phone screen, or to activate object avoidance maneuvers, requires significant processing power. For example, the DJI Phantom 4 introduced a significant development in the evolution of civilian drones by employing a dedicated visual processing chip created by Dublin-based start-up Movidius – the Myriad 2 Visual Processing Unit. This is a specialized chip for high definition and real-time visual processing at ultralow power, which means it improves the capabilities of drones, their mobility, maneuverability and autonomy.

The Myriad 2 chip is designed specifically for processing machine vision and integrating data from multiple sensors to allow the device to make sense of surroundings, avoid obstacles or track and follow objects. In a promotional video for Movidius and DJI, Remi El-Ouazzane explains his company's hardware development and DJI's machine vision algorithms in human terms: "Visual intelligence is about giving the power of sight to our devices," a generational step that moves drones from capturing high-resolution aerial imagery to having "the ability to understand what they are looking at"; we are entering, he notes, "the golden age of embedded computer vision."[90] This is not an overstatement of the shift at play in the deployment of high-powered visual processing chips in mobile devices. Not insignificantly, this step empowers personal mobile devices with the same spatial awareness, depth perception and visual analysis that guide drones.

When considering the impact of mobiles on society, on communication and personal relationships, it is easy to overlook these transformations in image

FIGURE 4.1 Mobile media, augmented reality, Project Tango. Image: Maurizio Pesce.

Source: https://commons.wikimedia.org/wiki/File:Project_Tango_(24181870520).jpg

processing. And, when we pair the capacity for imaging and image processing with *mobility*, a wider set of devices come into alignment with camera phones and smartphones, including GoPro cameras, wearable cameras (Google Glass and its variants, such as Narrative Clip)[91] and 360-degree cameras, for instance. When GoPro is granted a new patent for "image sensor data compression and DSP decompression,"[92] among a range of other image processing tech patents, including 360-degree imaging, we are seeing the expansion of the set of net-worked mobile imaging devices well beyond the phone. It makes sense that GoPro extended its focus on sports cameras to building its own drones, as the platform on which to mount and bring new mobility to camera devices. The image processing patents are a means for connecting that imaging to networked distribution and to cloud-based image processing.

In addition to powering drones, the Myriad 2 chip was also integrated into Google's Project Tango in 2014 and into Tango-enabled phones through part-nership with Lenovo in 2016 (Figure 4.1).[93]

Project Tango: To Map an Interior Is to Infinitely Expand Our Visible Publics

What started its life as "Project Tango," and is now simply Tango, to an extent hides its own social innovation. It perfectly illustrates the phone as a platform for generative imaging and automated sensor-based mapping as a wholesale geographical information system (GIS). In other words, before it became explicitly framed as an AR platform, it was touted as a key development in the quest for the holy grail of interior mapping and accurate mobile spatial awareness – a quest that is now pursued with gusto by others, including indoor-cartography specialists, Micello, who work with shopping centers, airports and the like.

Google, however, has since widened its ambit to include more consumer-friendly applications in AR, location-based gaming, real-world search and so on. The last thing the company would have wanted was public scrutiny of its system's ability to map, know and understand the interior spaces of everywhere we go, accurately monitor and remember where we've been and what was there and perhaps predict where we're going before we get there. We can imagine a society of walking sensor "drones" ceaselessly mapping the internal spaces that are out of reach to Street View cars with their mounted 360 cameras. But if its use is in realizing the vast possibilities of AR applications, the technology that facilitates this is firmly situated within the new camera consciousness.

Tango-enabled mobile devices reconfigure the mediated interiors of shops, galleries, homes or any other space. They enable dynamic, personalized gaming environments, where one's own place unfolds as the game map while it is traversed. Spatial and object awareness aids the development of assistive technology applications,[94] including internal navigation for visually impaired people. They automate and distribute mobile points of view like never before, and further detach imaging from individuals. Commercial and promotional uses are shaping some of the more advanced applications of AR so far. Automotive giant BMW has used the Tango AR platform to launch its i3 and i8 cars, no doubt as a marketing ploy, allowing app users on Tango-enabled devices to view a lifelike 3D model against real-world backdrops, as well as "enter" the cars and explore their interior.[95]

Tango can best be understood as an assemblage of hardware, sensors and software that responds to user movement, positioning and activity with a high degree of accuracy. In this sense, it also acts as a media platform upon which applications can be built.

Google has built Tango as an integrated array of sensors, chip hardware – including the Myriad 2 visual processing unit, Intel and Qualcomm chips for high-speed processing and coordination of sensing and communication components – and software involving cloud-based visual processing (to lower the power drain on devices). There are three core elements to the technology: positioning and orientation, depth perception through a RealSense 3D camera offering 3D object rendering and area learning – which is where the device maps and remembers its surroundings. Its motion-tracking sensors combine visual features of the environment with an accelerometer and gyroscope to track the device's movements in space. These are mapped as "poses," which are about capturing motion, area learning and depth perception. The camera is key and, for instance, will calibrate motion through, among other ways, visual-inertial odometry (VIO), which estimates "where a device is relative to where it started."[96]

Tango-enabled devices make sense of their environment in the form of events, which are the mapping of "the pose (position and orientation) of the device," "frames and textures from a camera" and "point clouds, which are generated via depth sensing."[97] Ultimately, the device becomes a communication platform that knows its own environment at every moment, remembers "areas that it has travelled through and localizes the user within those areas up to an accuracy of a few centimeters."[98] In Google's own words:

> Tango lets you see more of your world. Just hold up your phone, and watch as virtual objects and information appear on top of your surroundings. So, no matter where you are, there's always a richer, deeper experience to engage with, explore and enjoy. You'll see.[99]

Google's promotional image (Figure 4.2) presents a peculiar vision of its visual and spatial awareness technology. It aesthetically transforms and overlays all aspects of the scene with vector maps and nonhuman visual spectrum. In this account, the phone is disembodied, or the body becomes obsolete to the operations of the device in rendering its environment. However, in practice, this kind of technology actually delivers something quite the opposite. It materializes personal media and automates geospatial communication to a greater degree than anything before it. The "pose" of the hand and phone provides vital data. Positioning in space and relationships between the subtle movements of the

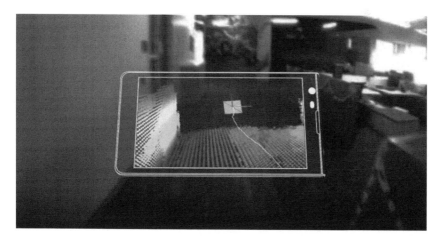

FIGURE 4.2 Google's Project Tango, interior mapping and spatial awareness technology.
Source: Google.

device and objects in the environment establish an all-important set of dynamic relations that create area awareness, and its functionality, for AR applications.

Rather than a single technological step, computer vision and digital image processing are components in a much larger assemblage of software, hardware, protocols, user practices, norms of device use and so on, which have to be brought together in a way that enables devices as "seeing machines" equipped with area awareness as a *platform* on which to build AR functionality. Google's SkyHook Wireless patent dispute was about standardization across different services and avoiding data contamination and dissonance, and hence securing the means to the platform on which developers can layer applications and functionality.[100] At Google's 2017 developer conference, I/O 2017, they announced a further refinement to Tango, dubbed the "Visual Position Service." This has been described as "a collaboration with Google Maps where indoor environments are mapped out and tagged within the view of a mobile phone camera."[101]

Again, many of these developments, as we have been at pains to argue in this chapter, have not been without precedents or parallel developments. Rudimentary indoor mapping using photographic imagery, for instance, has long been possible using photogrammetry, which is a three-dimensional coordinate measuring technique that uses photographs, and triangulation, as "the fundamental medium for metrology (or measurement)."[102] In addition, the tagging of information within the view of a mobile phone camera is not exclusive to Tango, with Amsterdam-based Layar (founded in 2009, but subsequently sold and since closed) proving influential with its development of an early mobile AR service[103] – one, like Tango, that worked to transform "visual universes into informational universes."[104]

What is particularly striking about Google, though, has been its ability to incorporate Tango into its other services. For example, in addition to Tango, Google has also announced a new smartphone image recognition feature called Lens – which morphed from an earlier Google product, Google Goggles[105] – that permits the phone camera to recognize complicated images and identify and act on information contained within them, such as providing full business information of a restaurant when the camera is aimed at shop signage from across the street.[106] Google have brought this assemblage of technologies of visibility together in Tango-enabled mobile phones as a way of embedding vision capture into everyday life. Google's goal in doing so is to provide its platform as a whole technological base for integrating geospatial data into Internet-based and cloud-processed applications. And this goal has broad-reaching implications for the social.

The Changing Terrain of Visibility

Recent approaches to critical geographic information systems (GIS) emphasize the way that this kind of geospatial and mapping technology "is not just software, but actually produces society."[107] GIS refers both to the technologies or software systems that allow complicated mapping involving layered and often dynamic real-time data, as well as the field of study of geospatial mapping and its social implications. For Wilson, GIS "have become so central to daily life in advanced capitalist societies as to become invisible, part of our technological unconscious."[108] GIS, in other words, organizes society logistically, but in ways that some argue are or should be contested. For instance, critical GIS raises concerns about the openness of location and geospatial data and mapping, whether it forms the exclusive property of an Internet giant like Google or becomes an open, common public source from which to generate a broad range of knowledge, not just those of commercial value or tied solely to consumption and marketing.[109] As with the shift from image to imaging, GIS deals with and creates the conditions for shifting from map formation to dynamic and interactive mapping.

Many examples of dynamic, interactive geographic information systems have been produced over the last decade, drawing on mobile and social media data overlaid and layered with geospatial and locational data. Projects like OpenStreetMap, founded in 2006, or crisis mapping tools like Ushahidi, founded in 2008, rely on mobile devices and GPS data for user-generated input. Geotagged data has been extracted from popular social media platforms such as Twitter and Instagram for alternative purposes. Jeremy Crampton and colleagues argue for the need to move beyond the simple geotag as a "collection of latitude-longitude coordinates extraneously attached to other bits of information, and instead understand it as a socially produced space that blurs the oft-reproduced binary of virtual and material spaces."[110] And, there are numerous projects and services that conform to this understanding. CityBeat, for instance, is a project developed by researchers at the Jacobs Institute at Cornell Tech, The New York World (Columbia Journalism School), Rutgers University, NYU, and Columbia University. CityBeat draws together and analyzes geotagged social media data, including visual content from Twitter and Instagram, layered to urban maps.[111] Algorithms look for anomalies or intensities to highlight activities of interest as they are taking place.

If one of the endpoints of GIS developments and the area awareness capabilities of Tango is the goal of predictive analytics, then Banjo (ban.jo) stands as one of the more ambitious platforms, in that it has attempted to provide a service that allows subscribers to "see and know things happening in the world before anybody else," as the company's founder Damien Patton puts it. Talking about the company's turn toward predictive image analytics, around the time of the Boston Marathon bombing in 2013, Patton explains: "If we can automate, build AI in a system that literally would have known this in real time . . . ahead of time, imagine how that would have changed everything. There would have had to be no human involvement."[112] The company stores information from a wide range of social media services and other public data sources. It uses these historical data so as to be able to analyze disruptions to norms and to identify anomalies. By making sense out of disparate signals, in the form of structured and unstructured visual data, Banjo's intervention is not simply about compiling or mapping images in time and place, but enabling a mechanism for paying attention and visibilizing events and activities. AI becomes essential in the attempt to build camera conscious systems able to act and intervene while mapping and sensing local environments.

Image acquisition and sensing capabilities of mobile smartphones and their internal visual processing chips and arrays of sensors follow the rapid growth of personal and social functionality for mobile devices. What we are highlighting here are some significant ways mobile devices are becoming integrated into sensing and mapping platforms, and incorporating people into the process in ways that bring new dimensions to our conscious and unconscious imaging and visibility practices. Across the examples we have explored here – from Google's Tango's augmented reality to Banjo's real-time social media image analysis – the significance, and wider potential, of these developments rests in their capacity to combine varied technologies of visibility and mobility (internal mapping and "area awareness," aspects of accelerometer and other means of registering position and movement). Through the combination and overlay of space and time, movement and activity, these intimate technologies of visibility enable the creation of detailed portraits of how human movement is entwined with the complexities and banalities of everyday life, with traces, threads, passages and lines,[113] spikes and intensities and fluctuating modes of engagement (trends and anomalous patterns or "off-trends").[114] The urban environment thus mapped and made visible in new ways can also be understood in relation to, and integrated within, another, emergent, form of "seeing machine": autonomous vehicles. Before addressing the machine vision at work in autonomous vehicles, we turn to an examination of drones and the phenomenon of calculated vision, recorded from above.

Notes

1. James E. Katz and Mark A. Aakhus, eds., *Perpetual Contact: Mobile Communication, Private Talk, Public Performance* (Cambridge: Cambridge University Press, 2002); Rich Ling, *The Mobile Connection: The Cell Phone's Impact on Society* (San Francisco, CA: Morgan Kaufmann, 2004); Gerard Goggin, *Cell Phone Culture* (London: Routledge, 2006).
2. Jane Bennett, *Vibrant Matter. A Political Ecology of Things* (Durham, NC: Duke University Press, 2010), xii.
3. We use the term "social imaging" to characterize the observations made by José van Dijck (2007) and Rubenstein and Sluis (2008), among others, that personal image making has shifted from memory capture and archiving to visual communication and sociality, particularly in their integration with social media platforms. See: José van Dijck, *Mediated Memories in the Digital Age* (Stanford, CA: Stanford University Press, 2007); Daniel Rubenstein and Katrina Sluis, "A Life More Photographic: Mapping the Networked Image," *Photographies* 1, no. 1 (2008): 9–28.

4. Tim Bradshaw, "How Apple's Ultra-wideband Chip Could Transform Its Products," *Financial Times*, September 20, 2019, www.ft.com/content/47e914a0-da3b-11e9-8f9b-77216ebe1f17.
5. See, for example: Eric Limer, "Microsoft's New Real-World Search Engine Is Incredible and Horrifying," *Popular Mechanics*, May 10, 2017, www.popularmechanics.com/technology/infrastructure/news/a26456/microsoft-machine-learning-vision/.
6. Rich Ling, *New Tech, New Ties: How Mobile Communication Is Reshaping Social Cohesion* (Cambridge, MA: MIT Press, 2008); Gerard Goggin, *Global Mobile Media* (London: Routledge, 2011); Rowan Wilken and Gerard Goggin, eds., *Mobile Technology and Place* (New York: Routledge, 2012); Leighton Evans and Michael Saker, *Location-based Social Media: Space, Time and Identity* (Houndmills, Basingstoke: Palgrave Macmillan, 2017); Jason Farman, *Mobile Interface Theory: Embodied Space and Locative Media* (New York: Routledge, 2012); Adriana de Souza e Silva and Jordan Frith, *Net Locality: Why Location Matters in a Networked World* (London: Wiley, 2011); Jordan Frith, *Smartphones as Locative Media* (Cambridge: Polity Press, 2015).
7. Nana Verhoeff, *Mobile Screens: The Visual Regime of Navigation* (Amsterdam: Amsterdam University Press, 2012).
8. Daniel Palmer, "Redundant Photographs: Cameras, Software and Human Obsolescence," in *On the Verge of Photography*, Daniel Rubenstein, Johnny Goldin, and Andy Fisher (Eds.) (Birmingham: ARTicle Press, 2013), 49.
9. Chris Chesher, "Between Image and Information: The iPhone Camera in the History of Photography," in *Studying Mobile Media: Cultural Technologies, Mobile Communication, and the iPhone*, Larissa Hjorth, Jean Burgess, and Ingrid Richardson (Eds.) (New York: Routledge, 2012), 101.
10. Palmer, "Redundant Photographs," 50.
11. William Eureka, "Simplify Before You Automate," *Material Handling & Logistics*, April 1, 2010, www.mhlnews.com/technology-amp-automation/simplify-you-automate#close-olyticsmodal.
12. Todd Gustuvson, *Camera: A History of Photography from Daguerreotype to Digital* (New York: Sterling, 2009), 320–321.
13. Reese V. Jenkins, "Technology and the Market: George Eastman and the Origins of Mass Amateur Photography," *Technology and Culture* 16, no. 1 (1975): 16.
14. Ibid., 16.
15. Chesher, "Between Image and Information," 102.
16. Ibid., 148.
17. Gustuvson, *Camera*, 234.
18. Ibid., 234.
19. Ibid., 298.
20. Ibid.
21. Ibid., 302.
22. Ibid.
23. Ibid., 300.
24. Ibid., 312.
25. Ibid., 315.
26. Ibid.
27. Amr Mohamed Galal, "An Analytical Study on the Modern History of Digital Photography," *International Design Journal* 6, no. 2 (2016): 205.
28. Ibid., 205.
29. Ibid.
30. Ibid., 206.
31. Ibid.
32. For detailed discussion, see ibid., 206–215.
33. John Berger, *Understanding a Photograph* (New York: Aperture, 2013), 50.
34. Palmer, "Redundant Photographs," 50.

35. Estelle Blaschke, "From Microform to the Drawing Bot: The Photographic Image as Data," *Grey Room* 75 (2019): 65.

36. Ibid., 60–83.

37. Ibid., 79.

38. Daniel Palmer, "Photography as Indexical Data: Hans Eijkelboom and Pattern Recognition Algorithms," in *Photography and Ontology: Unsettling Images*, Donna West Brett and Natalya Lusty (Eds.) (New York: Routledge, 2019), 132. See also: Daniel Palmer, "The Rhetoric of the JPEG," in *The Photographic Image in Digital Culture*, 2nd ed, Martin Lister (Ed.) (London: Routledge, 2013), 149–164.

39. Blaschke, "From Microform," 65.

40. Ibid., 79.

41. Chesher, "Between Image and Information," 105.

42. Ibid.

43. Ibid., On the history of CCD (charged coupled device) and CMOS (complementary metal-oxide semiconductor) image sensors and their use in mobile phones, see: Sean Cubitt, *The Practice of Light: A Genealogy of Visual Technologies from Prints to Pixels* (Cambridge, MA: MIT Press, 2014), 100–111.

44. Chesher, "Between Image and Information," 105.

45. Mikko Villi, "Visual Mobile Communication on the Internet: Patterns in Publishing and Messaging Camera Phone Photographs," in *Mobile Media Practices, Presence and Politics: The Challenge of Being Seamlessly Mobile*, Kathleen M. Cumiskey and Larissa Hjorth (Eds.) (New York: Routledge, 2013), 217.

46. Martin Lister, "Is the Camera an Extension of the Photographer?" in *Digital Photography and Everyday Life*, Edgar Gómez Cruz and Asko Lehmuskallio (Eds.) (New York: Routledge, 2016), 213.

47. Fumitoshi Kato, Daisuke Okabe, Mizuko Ito, and Ryuhei Uemoto, "Uses and Possibilities of the Keitai Camera," in *Personal, Portable, Pedestrian, Mobile Phones in Japanese Life*, Mizuko Ito, Misa Matsuda, and Daisuke Okabe (Eds.) (Cambridge, MA: MIT Press, 2005), 300–310.

48. Ilpo Koskinen, "Seeing with Mobile Images: Towards the Perpetual Visual Contact," in *Proceedings of the Conference The Global and the Local in Communication: Places, Images, People, Connections*, Budapest, June 10–12 June, 2004; Fumitoshi Kato, "Seeing Seeing of Others: Conducting a Field Study with Mobile Phones/Camera Phones," in *Proceedings of the Conference Seeing, Understanding, Learning in the Mobile Age*, Budapest, June 10–12, 2004.

49. Larissa Hjorth and Sarah Pink, "New Visualities and the Digital Wayfarer: Reconceptualizing Camera Phone Photography and Locative Media," *Mobile Media and Communication* 2 no. 1 (2014): 40–57.

50. Mikko Villi, "Visual Chitchat: The Use of Camera Phones in Visual Interpersonal Communication," *Interactions: Studies in Communication and Culture* 3, no. 1 (2012): 39–54.

51. Amparo Lasén and Edgar Gómez-Cruz, "Digital Photography and Picture Sharing: Redefining the Public/Private Divide," *Knowledge, Technology and Policy*, no. 22 (2009): 205–215.

52. Katharina Lobinger, "Photographs as Things – Photographs of Things: A Texto-Material Perspective on Photo-Sharing Practices," *Information, Communication & Society* 19, no. 4 (2016): 475–488.

53. Daniel Rubinstein and Katrina Sluis, "A Life More Photographic," *Photographies* 1, no. 1 (2008): 9–28.

54. Susan Murray, "Digital Images, Photo-Sharing, and Our Shifting Notions of Everyday Aesthetics," *Journal of Visual Culture* 7, no. 2 (2008): 147–163.

55. Chris Peters and Stuart Allan, "Everyday Imagery: Users' Reflections on Smartphone Cameras and Communication," *Convergence: The International Journal of Research into New*

Media Technologies 24, no. 4 (2018): 357–373. See also: Lisa Gye, "Picture This: The Impact of Mobile Camera Phones on Personal Photographic Practices," in *Mobile Phone Cultures*, Gerard Goggin (Ed.) (New York: Routledge, 2008), 135–144; and Nancy A. Van House, "Collocated Photo Sharing, Story-telling, and the Performance of Self," *International Journal of Human-Computer Studies* 67 (2009): 1073–1086.

56. Heidi Rae Cooley, "It's All About the *Fit*: The Hand, the Mobile Screenic Device and Tactile Vision," *Journal of Visual Culture* 3, no. 2 (2004): 145.

57. Ingrid Richardson and Rowan Wilken, "Parerga of the Third Screen: Mobile Media, Place, and Presence," in *Mobile Technology and Place*, Rowan Wilken and Gerard Goggin (Eds.) (New York: Routledge, 2012), 181–197.

58. Spencer Shaw, *Film Consciousness: From Phenomenology to Deleuze* (Jefferson, NC: McFarland, 2008).

59. Andrea Mubi Brighenti, *Visibility in Social Theory and Social Research* (Houndmills, Basingstoke: Palgrave Macmillan, 2010).

60. Ibid.

61. Ibid., 37–38.

62. Ibid., 37.

63. Ibid.

64. Adrian Mackenzie, *Wirelessness: Radical Empiricism in Network Cultures* (Cambridge, MA: MIT Press, 2010), 14.

65. Brian Merchant, *The One Device: The Secret History of the iPhone* (London: Bantam Press, 2017); Pelle Snickars and Patrick Vonderau, eds., *Moving Data: The iPhone and the Future of Media* (New York: Columbia University Press, 2012); Larissa Hjorth, Jean Burgess, and Ingrid Richardson, eds., *Studying Mobile Media: Cultural Technologies, Mobile Communication, and the iPhone* (New York: Routledge, 2012).

66. Chesher, "Between Image and Information," 106.

67. Daniel Palmer, "iPhone Photography: Mediating Visions of Social Space," in *Studying Mobile Media: Cultural Technologies, Mobile Communication, and the iPhone*, Larissa Hjorth, Jean Burgess, and Ingrid Richardson (Eds.) (New York: Routledge, 2012), 87.

68. Ibid., 85.

69. Ibid., 88.

70. Ibid.

71. Ibid.

72. Chesher, "Between Image and Information," 98.

73. Megan Halpern and Lee Humphreys, "iPhoneography as an Emergent Art World," *New Media & Society* 18, no. 1 (2016): 62–81.

74. Villi, "Visual Mobile Communication," 215.

75. Palmer, "iPhone Photography," 88.

76. Ibid.

77. Ibid.

78. Chesher, "Between Image and Information," 107.

79. Blaschke, "From Microform," 64.

80. Palmer, "iPhone Photography," 90.

81. Mark Andrejevic, "Becoming Drones: Smartphone Probes and Distributed Sensing," in *Locative Media*, Rowan Wilken and Gerard Goggin (Eds.) (New York: Routledge, 2015), 193–207.

82. Palmer, "iPhone Photography," 86–87.

83. Brian Hayes, "Computational Photography," *American Scientist* 96, no. 2 (2008), www.americanscientist.org/article/computational-photography. The idea of the smartphone camera as a tool that is making (rather than just taking) images is most apparent in the more recent "night mode" features of Android and iOS devices. When this feature is used, the final image is a result of "bracketing,"

whereby the device will record a series of images at different exposure levels, and then merge them together into a single image using specialized software. See: Edgar Cervantes, "What Is Night Mode and How Does It Work?" *Android Authority*, April 28, 2019, www.androidauthority.com/what-is-night-mode-and-how-does-it-work-979590/.

84. Palmer, "iPhone Photography," 50, 90. See also: John Tagg, "Mindless Photography," in *Photography: Theoretical Snapshots*, J. J. Long, Andrea Noble, and Edward Welch (Eds.) (London: Routledge, 2009), 16–30.

85. Trevor Paglen, "Invisible Images (Your Pictures Are Looking At You)," *The New Inquiry*, December 8, 2016, https: thenewinquiry.com/invisible-images-your-pictures-are-looking-at-you/.

86. David Bate, "The Digital Condition of Photography: Cameras, Computers and Display," in *The Photographic Image in Digital Culture*, 2nd ed., Martin Lister (Ed.) (New York: Routledge, 2013), 79.

87. Bate, "The Digital Condition of Photography," 79.

88. Lister, "Is the Camera," 213.

89. Palmer, "iPhone Photography," 94.

90. Remi El-Ouazzane, Movidius CEO, quoted in "Movidius and DJI Bring Vision-Based Autonomy to Phantom 4," YouTube video, 3:28. "Intel Movidius," March 16, 2016. https://youtu.be/hX0UELNRR1I?list=PL0PjXgwbk-vur0wGByQSV u9r1ODn5TK8m.

91. See: Jill Walker Rettberg, *Seeing Ourselves Through Technology: How We Use Selfies, Blogs and Wearable Devices to See and Shape Ourselves* (Houndmills, Basingstoke: Palgrave Macmillan, 2014).

92. Gopro, Inc. "Image Sensor Data Compression and DSP Decompression," US Patent Office, US20160227068 A1, August 4, 2016, www.google.com/patents/US20 160227068.

93. Ina Fried, "Tiny Chipmaker Movidius Has a Tiny Chip That Looms Large," *Recode*, March 16, 2016, www.recode.net/2016/3/16/11587020/tiny-chipmaker-movidius-has-a-tiny-chip-that-you-will-be-seeing-a-lot. In related developments, San Francisco-based firm 6D.ai is working with Qualcomm Technologies to provide technology for Qualcomm's Snapdragon mobile processing platform, with the aim of building "a 3D map of the world using only smartphone cameras." See: Dean Takahashi, "6D.ai Creates Platform for a 3D Map of the World," *Venture Beat*, August 27, 2019, https://venturebeat.com/2019/08/27/6d-ai-creates-platform-for-a-3d-map-of-the-world/.

94. See: Rabia Jafri and Marwa Mahmound Khan, "Obstacle Detection and Avoidance for the Visually Impaired in Indoor Environments Using Google's Project Tango Device," in *Computers Helping People with Special Needs. ICCHP 2016. Lecture Notes in Computer Science, vol 9759*, K. Miesenberger, C. Bühler, and P. Penaz (Eds.) (Cham: Springer, 2016), 179–185, https://doi.org/10.1007/978-3-319-41267-2_24; Rabia Jafri, et al., "Utilizing the Google Project Tango Tablet Development Kit and the Unity Engine for Image and Infrared Data-Based Obstacle Detection for the Visually Impaired," *International Conference on Health Informatics and Medical Systems | HIMS'16, 2016*, 163–164, https://pdfs.semanticscholar.org/6572/49b121859df27b d8b06787b7e9df4335663b.pdf.

95. Sam Shead, "BMW Hopes Google's Augmented Reality Tango Technology Will Help It Sell Cars," *Business Insider*, January 5, 2017, www.businessinsider.com/bmw-google-augmented-reality-tango-2017-1?r=US&IR=T&IR=T.

96. Tango, "Motion Tracking," https://developers.google.com/tango/overview/motion-tracking.

97. Tango, "Overview: Events," https://developers.google.com/tango/overview/events.

98. Jafri, et al., "Utilizing the Google Project Tango Tablet Development Kit."

99. See: https://get.google.com/tango/.
100. Nilay Patel, "How Google Controls Android: Digging Deep into the Skyhook Filings," *The Verge*, May 12, 2011, www.theverge.com/2011/05/12/google-android-skyhook-lawsuit-motorola-samsung.
101. Robert Mueller, "Google Adds New Tricks and Devices to Tango Augmented Reality," *Fast Company*, May 17, 2017, https://news.fastcompany.com/google-adds-new-tricks-and-devices-to-tango-augmented-reality-4038110.
102. Anil Narendran Pillai, "A Brief Introduction to Photogrammetry and Remote Sensing," *GIS Lounge*, July 12, 2015, www.gislounge.com/a-brief-introduction-to-photogrammetry-and-remote-sensing/.
103. On Layar, see: Tony Liao and Lee Humphreys, "Layar-ed Places: Using Mobile Augmented Reality to Tactically Reengage, Reproduce, and Reappropriate Public Space," *New Media & Society* 17, no. 9 (2015): 1418–1435; Nanna Verhoeff, "A Logic of Layers: Indexicality of iPhone Navigation in Augmented Reality," in *Studying Mobile Media: Cultural Technologies, Mobile Communication, and the iPhone*, Larissa Hjorth, Jean Burgess, and Ingrid Richardson (Eds.) (New York: Routledge, 2012), 118–132.
104. Chesher, "Between Image and Information," 111.
105. See: Chesher, "Between Image and Information," 110–111; C. Scott Brown, "Google Updates Goggles After 3 Years Just to Tell People to Install Lens Instead," *Android Authority*, August 17, 2018, www.androidauthority.com/google-goggles-lens-896203/.
106. Mark Sullivan, "Google Announced 'Lens' for Advanced Image Recognition in Smartphones," *Fast Company*, May 17, 2017, https://news.fastcompany.com/google-announced-lens-tool-for-advanced-image-recognition-in-smartphones-4038050.
107. Matthew W. Wilson, "Critical GIS," in *Key Methods in Geography*, 3rd ed., Nicholas Clifford, Shaun French, and Gill Valentine (Eds.) (London: Sage, 2016), 285–301.
108. Wilson, "Critical GIS," 288. On the technological unconscious, see: Nigel Thrift, "Remembering the Technological Unconscious by Foregrounding Knowledges of Position," *Environment and Planning D: Society and Space* 22 (2004): 175–190.
109. Agnieszka Leszczynski, "Situating the Geoweb in Political Economy," *Progress in Human Geography* 36, no. 1 (2012): 72–89.
110. Jeremy W. Crampton, Mark Graham, Ate Poorthuis, Taylor Shelton, Monica Stephens, Matthew W. Wilson, and Matthew Zook, "Beyond the Geotag: Situating 'Big Data' and Leveraging the Potential of the Geoweb," *Cartographic and Geographic Information Science* 40, no. 2 (2013): 132.
111. See: CityBeat, http://thecitybeat.org/.
112. Retrieved from https://ban.jo.
113. Tim Ingold, *Lines: A Brief History* (London: Routledge, 2007), 52.
114. Max Sklar, Blake Shaw, and Andrew Hogue, "Recommending Interesting Events in Real-time with Foursquare Check-ins," in *Proceedings of the Sixth ACM Conference on Recommender Systems, September 9–13, 2012, Dublin, Ireland*, www.metablake.com/foursquare/recsys2012.pdf.

5

DRONE VISION

If we look at what are commonly called drones, there is one feature we see in almost every situation – the presence of a camera.[1]

In 1969, William H. Whyte was commissioned to study New York's urban spaces in what became an influential approach to understanding and improving the impacts of urban design. The Street Life Project used simple but innovative observational techniques combining an urban ethnography with data extracted from time-lapse cameras placed on rooftops overlooking 14 plazas and three small parks. Informed in part by this calculated view from above, Whyte describes the more sociable plazas he studied as highly visual environments characterized by people looking at other people, the interplay of stasis and movement, and interactivity with the material features of the public spaces.[2] Whyte's "unobtrusive" observations and his use of topological perspective to map the movement, groupings and interactions between people in public space were significant innovations for urban planning and human geography.

The use of camera analysis in Whyte's studies also has an interesting resonance with the way aerial drones have emerged in cities around the world, including in times of protest, to reveal and re-vision urban spaces. An early example of the use of drone vision in protest zones occurred during Poland's 2011 Independence Day rallies. The cavernous, gridded streets of Warsaw became the site of running battles between Nationalists and anti-fascists, with riot police attempting to keep the two groups apart. Drone footage originally attributed to YouTube user latajacakamera circulated showing a particular kind of intervention into the scene.[3] The rudimentary and seemingly homemade drone ascends from behind one group of protesters and provides a study of the scene from above. Reminiscent of Whyte's fixed aerial visual study of New York's public spaces, but this time on the move and traversing a more dramatic space, the drone camera crosses the police lines, providing panoramic views of Warsaw's besieged streets and revealing the strong and weak lines of riot police and protest groups. The drone follows groups of officers running to fill gaps in their lines, and takes in a city struggling to maintain its designed public order. In November 2017, more advanced, lightweight drones with high resolution, sophisticated cameras rose high above the same streets and squares to capture iconic images of red-flared mass of ultra-right and neo-Nazi protesters. Again, the aerial camera is able to establish and track the extent of the protests and their interactions and confrontations with others – counter protesters, police – within the city.

Tensions raised by aerial vehicles within city centers, and their view from above, were brought to the surface in their use in Turkey amidst ongoing clashes between a protesting public and the ruling party. In Istanbul's Taksim Square and Gezi Park, 2013, the embattled Turkish government under Recep Teyyip Erdogan committed military and police forces to ongoing battles with protesters. One protester flew camera-mounted drones for days, offering real-time aerial vision of the more chaotic and disorganized police movements, the reach of water cannon from anti-riot vehicles and areas engulfed in teargas. Around the bodies on the ground, the water cannon vehicles and the drone form an assemblage of action and counteraction resulting, finally, in the drone being shot down by police.[4] In this instance, the visual study of the square is accompanied by an explicitly camera conscious crowd. Both sides of the contest have an investment in the drone's vision, but on this occasion its impact lies heavier on the police who were trying, often unsuccessfully, to control movement and spatial occupation in the square. As the transmission of drone vision is severed, there is a disjunctive return of the optical field to a ground-level line of sight. But, not surprisingly, this moment was also captured by camera phone footage from below.[5]

Around the world, similar protest sites have been reshaped by the presence of civilian drones. The so-called umbrella movement that emerged during Hong Kong's 2014 democracy protests took its name from the use of umbrellas as a symbol of protest and city occupation. In a city already known for its high-rise verticality, images of tens of thousands of open umbrellas at street level were made most visible from above through extensive drone vision. The canopy of umbrellas made most sense visually as drones wove through the buildings above the massed protesters.[6] Umbrellas took on an additional role during the 2019 pro-democracy, anti-extradition law protests. Instead of the symbolism of an umbrella canopy, they served as a shield to protect protesters from suspected face recognition cameras positioned above them throughout the city. Similarly, the Bangkok shutdown in January 2014 was monitored by drone vision, making visible the extent of the protest but also showing military leaders where "hotspots" were throughout the city.[7] In each of these cases, there is a clearly expanded awareness of the scene of protest, of the extent and power in the image of masses in public spaces. The drone camera inhabits this aerial space, acting as an unbounded proxy for those contesting the streets below and affecting the visual contours of the protesting public. In this sense, it belongs to a camera conscious protest crowd aware of the implications of altering public visibility by use of the drone camera.

These events exemplify the actions of drones in making visible what has remained hidden in most popular and academic accounts of drones: the image of the aerial camera in its unfixed, "motile," semiautonomous re-visioning of the social.[8] These instances of drone use illustrate their ability to alter and automate aspects of seeing, to "amass" groups and congregations, as well as pinpoint people above an urban environment. Like Whyte's innovative observational research methodology, drones are able to map human movement and collectives in significant new ways. For many, these capabilities only add to the threat that new camera technologies pose. Carolyn Marvin was writing about mobile media and camera phones when she noted that "when new forms put familiar social distances at risk, terrible anxieties ripple through the body politic."[9] She used the notion of "asymmetrical transparency" to invoke the cause of this anxiety, and this notion certainly relates to the new camera consciousness that drones bring into being. There are, however, other aspects to drone vision worth our attention.

This chapter adds a counterpoint to Marvin's "terrible anxieties." We look to drones (in all their varied shapes, sizes and functionality but with a focus on small civilian aerial drones) as they effect vision and visibility, while maintaining focus on the ethical implications of the power relations they establish or skew. In addition, considering what a drone sees, there is a need to better understand drone vision from below, through the human connection as a form of vision augmentation. This link illustrates the new modes of visual data capture and forms the basis for drones' socio-technical applications in monitoring, mapping, inspecting and analyzing, and, ultimately, as they intervene in geographical space and social relations. These semisubjective, mobile-but-tethered functions of the drone camera offer a new kind of "geomedia" in the creation of recompilable and calculable public space. Our argument is that this special case of geomedia relies on an integrated ecosystem of urban (and aerial) spatial management, wireless infrastructures, automation in the robotics and visual sensing and first-person view (FPV) as these enable drone flight as both locus of device control and data collection capability but also the source of drones' strangeness. Any account of aerial drones is incomplete without considering these augmentations as they reconfigure vision and visibility. In other words, we need to ask what kind of camera consciousness the aerial drone brings into being or signals through its capacity to generate distributed, augmented and semiautonomous vision from above.

Vertical Publics and Nonhuman Vision

Drone systems enter an already contested and carefully regulated vertical public space or orbital zone, which Lisa Parks considers as a neglected area of media research and public scrutiny.[10] The lack of critical attention to satellites indicates the success of the Western military-industrial-information complex at concealing their most strategic technologies in the process of applying global forms of informational and perceptual control. Regulatory conventions for governing vertical publics – spaces on the outer edges of vision and conceptualization, those orbital spaces and satellite trajectories – entail a relatively opaque politics.[11] Now, drones have emerged as a set of technologies that throw orbital power off its axis through their unfixed, unruly trajectories, their accessibility to ordinary users and their multidirectional motility.

The *motility* of drones, their autonomous or self-sustaining vertical and lateral movement, differentiates them from fixed surveillance devices or "the surveillant assemblage"[12] and the mobility of the personal mobile phone camera.[13] Their motility shifts the relations of object and subject, image and vision with attenuated and distributed modes of seeing and being seen. In part as a result of these shifts, national and international aviation bodies have scrambled to formulate regulation that balances drones' new capabilities for sensing environments and imagining landscapes against the wide-felt anxiety about the proliferation of drone cameras overhead.[14] Thus, the components of drone media create a volatile socio-technical and regulatory environment of object and signal-based relations, image relay and distribution.

In their maneuverability and agility, drones illustrate what William James called "a quasi-chaos."[15] James' point was that "there is vastly more discontinuity in the sum total of experiences than we commonly suppose."[16] Every movement, every new object within the environment complicates things further. In this sense, as each drone enters an airspace, that space becomes exponentially harder to control and at the same time that capacity for perceptual perspective or point of view fractures further. This has been the catalyst for fast-tracked drone regulation around the world, put in place by the same regulatory bodies that govern the movement of planes. The closure of Gatwick airport in the UK throughout December 2018 represented the first major limit case of the disruptive force of small personal drones on the basis of their unpredictability as a motile device.[17] But more than the physical presence of drones complicating the spatial arrangements of airspaces, the proliferation of aerial cameras greatly complicates the visual field.

Where satellites work on fixed vectors, providing rigid and expansive visual mapping from above, drone devices operate on directional freedom and targeted vision. And yet, what's equally important here is that, while satellites operate "on the perimeter of everyday visibilities and cultural theory,"[18] drones force an immediate and constant doubletake. Unlike Whyte's *inconspicuous* rooftop cameras, drones' noisy hovering presence is usually seen and felt. The technologies that enable drones to *see* and to *assess*, and hence to move independent of human control, also define their nonhuman agency, as they provoke, question, grapple and otherwise intervene in the spaces they navigate. While human control plays a large part in drone use, their systems increasingly rely on automated vision and movement, allowing constant micro-adjustments and object avoidance decision-making, in a powerful reconfiguration of the territory of the visible.

The relatively short history of satellite technology has illustrated the fact that the space stretching from earth to about 35,000 km above, from traversable airspace through low earth orbit (LEO) up to geostationary orbit (GEO), has become highly contested and inequitably regulated as a site of "international public affairs."[19] This space may be considered a "natural resource of advanced global civilization," as legal scholar Segfried Wiessner put it in 1983, during the height of the cold war; but it is one without "an equally advanced public order for its regulation."[20] Regulation of earth's vertical publics has been tied to the often competing interests of scientific innovation, national sovereignty and security, telecommunications infrastructure and the reconfiguration of the televisual. Inevitably, what may have been considered by Wiessner as a natural resource in 1983 has seen successive waves of colonization by wealthy nations and powerful global corporations. Opaque transnational treaties and governing bodies secure decisions far from public scrutiny about how access to the infrastructures of telecommunications and visuality from above is to be allocated and controlled. The value of this space, and hence contest over it, lies in the camera, wireless signal relay and the geo-locative information that satellites make possible. Consumer drones disrupt this scenario further in their resistance to regulatory control and, for now at least, in their DIY unpredictability.

The tumult that has followed these technological and social developments since the first satellite was launched in 1957 has quickened in recent years through the proliferation of drones. But it's not even very easy to pin down a generalizable definition of a drone. They are characterized as camera-mounted or remote sensing aircraft, known variously as Unmanned Aerial Vehicles (UAV) or Systems (UAS), or Remotely Piloted Aircraft (RPA). As Adam Rothstein points out: "We know what a drone is. But at the same time we don't."[21] Where "systems" are emphasized in the definitions of these vehicles, recognition is given to elements beyond the object – radio frequency spectrum, data link and video down link, "real-time" first person view (FPV), ground control systems, wireless local area networks and the chipsets and hardware that enable them. However, drones have entered the popular imaginary and regulatory discourse as a thing of risk and opportunity, an object of both fear and desire. The lack of standard nomenclature across regulatory bodies, manufacturers and user organizations flags the competing conceptual terrain in which drones operate.

One thing has remained consistent in the unfinished rise of this technology: "the presence of a camera."[22] Beyond the valid concerns about military uses of drones[23] and their effects on warfare and power, our interest lies in the role played by the introduction of intelligent vision as drones alter further the conditions of seeing and visibility. Like satellites before them, drones have moved beyond their military uses to reshape our vertical, visual publics, and force a rethinking of the territory of the knowable world. Drones have raised fears about catastrophic aerial accidents and heightening public concerns about optical surveillance and personal or commercial privacy invasion. These public contestations reveal something of the significance and social effects of the camera consciousness that follow drone technologies. They intensify certain features of the new camera consciousness in the palpable degrees of awareness of the agency and force of camera devices, provoking a reflexive understanding of an altered relationality that follows from being positioned before or in control of these mobile-networked cameras.

Drones, for the moment at least, interject by "making the camera felt," intensifying awareness of the properties of the camera in producing new modes of public visibility. They achieve this through their augmented modes of seeing as they move technically toward the increasing intelligence of the camera in its ability to visualize, calculate and process the visual field as visual data and act on those calculations. But this new camera consciousness is perhaps most evident in the value of the visual data drones are able to produce. For example, and in recognition of this, the Swedish government has required by law that anyone wishing to publish drone video and photography (on YouTube or Facebook, for instance) register with the Lantmäteriet, the authority responsible for protecting geographical data and information.[24]

A diverse range of practical uses are lined up against techno-failures and risks. A burgeoning industry in drone mapping for production, mining, land management and remote inspection work has developed. Agricultural uses align with delivery in automating a range of repetitive logistical tasks involving tracking, mapping and monitoring. New personal uses have arisen to take their nadir in the notion of the "dronie" – the use of drones in social media practices of self-imaging.[25] Autotracking systems based on visual recognition of a mobile subject accompany some drones to enable personal videography on the ski fields, for cyclists, surfers and other sporting activities. A new level of intelligence in machine vision allows aerial drones to operate autonomously. In this way, drone vision offers a vital functionality that alters the familiar relationality between cameras and operator in public spaces. In the same way that selfies can only be understood through the uses to which they are put within social networks,[26] drone vision is proliferating; but outside of its role in remote warfare, it is unsettled in its social functionality.

Successive waves of powerful and sophisticated civilian drones have introduced a shift in drone visuality from the operations and tropes of military surveillance and strikes and the sense of weaponized sight – Haraway's notion of the "God trick" of the surveillant gaze that remains beyond view, or the enlightenment roots of an objective "view from above."[27] But the camera consciousness brought into being by the drone camera transforms the surveillant gaze, blurring the balance between this *objective* view from above and the subjective point of view. The drone camera acts wirelessly, semiautonomously and with a liveliness that makes the camera felt. In these contexts, wirelessness, visual processing and visual signal and their politics come sharply into view, even if they are only felt as an awareness of the shift in the capacity for seeing at a distance and the sense of visibility.[28]

Object Relations, or the Creepy Thingness of Drones

The enigma of the drone lies not just in its precision remote flight capacities, but its reconfiguration of remote sensing, and the often-palpable disruption of the scene in which it enters (military, policing, search and rescue, landscape mapping, construction or public cinematography). By nature, drones dislocate and put in motion a camera operator's vision; but they also move vision beyond human perspective. Lisa Parks points to the pivotal role of infrared (IR) vision in military drones in generating "temperature data" or heat signals, "precisely so that it can be made productive within existing regimes of power."[29] This creates a potentially problematic "event-space," to use Virilio's term,[30] a relative and relational space that brings together heterogeneous human action, machine movement, perception and location. In the conflict zones of Northern Africa, Parks notes, those who are drawn into the drone's field of vision "become *spectral suspects* – visualizations of temperature data that take on the biophysical contours of a human body while its surface appearance remains invisible and its identity unknown,"[31] or, rather, is inferred as this data is cross-referenced with other signals, including mobile phone information.

It's important not to dismiss the labor and the emotional investment of the drone operator. Katherine Chandler demonstrates the "interpretive flexibility" that accompanied the development of military drones, a long process that began in the 1930s and a process that from the start saw troubled "divisions between human and nonhuman" in the way "unmanned" aircraft were imagined.[32] In her history of the development of drone technology, Chandler shows how "human and machine are produced in tension with each other," initially linked closely with the body and vision of the operator by military designers and champions, before "effacing" the controller so as to emphasize the "flying torpedo's" autonomy in the name of safety and superiority.[33] Even with personal or civilian drones, we have not fully come to grips with this tension. For Chandler, the disjuncture between drone and pilot had to be produced through the early developments that depicted drones as enhanced "unmanned" information gathering and weapon potentials – drones as a kind of ethical Kamikaze object. She shows how "there is no 'drone' that can be separated from the human operator, even if this disassociation was integral to the development of the weapon."[34]

Drones remain media devices that challenge our traditional views of the separation between the natural and the artificial. As the technology has become more sophisticated, smaller and seemingly more intelligent, these lines are blurred further still. Technical development has been driven by military goals. The Defense Advanced Research Projects Agency (DARPA), the agency responsible for laying the foundations for establishing the Internet, has backed extensive testing of the "swarming" capabilities and uses of drones. DARPA's explicit target is to find an advantage in battlefield visibility.[35] To overcome the constraints and limited sight lines of urban environments, drone swarms are designed to generate multiple points of view as an alternative saturated mode of automated visuality. Satellites and high-altitude reconnaissance drones can only provide a more or less two-dimensional vision of the target area, whereas drone swarms gain access to the variable topology of urban environments and the vertical spaces of buildings. The complex coordination of robotics is the secondary technology at play here; drone swarms actively reorient the capabilities of multi-viewpoint vision.

The ability of small insect-like drones to capture information, alongside the reconnaissance and "payload delivery" of military Reaper drones, has been imagined bluntly in the 2015 film *Eye in the Sky*.[36] In the film, a small-winged beetle is used to provide coveted internal vision of buildings that can't be viewed from the Reaper drones circling at high altitude above. What the film does explore in detail, though, is the extent of the information gathered by drones and the calculations made on the basis of that information. Screens streaming vision from the different drones to control rooms in the UK and the United States (where the drone operators are based) are filled with numerical data that gives semiotic validity to the visual data being "generated" by the drone. The aerial image builds a spatial blast diagram to provide a grim calculus of likely civilian fatality based on positioning against projected bomb radius. The film highlights the power embedded in the cameras' enhanced awareness of the urban setting, the regular and irregular movements of the Kenyan neighborhood under scrutiny, reminiscent of Whyte's Street Life Project, but with more serious consequences. Awareness generated by the cameras is contrasted by Kenyans' unawareness of them as they go about their daily activities.

While the fiction of the plausible beetle-shaped drone persists across the Internet, a number of less insectoid micro-drones are currently on the market. The US Army and other forces use a small, palm-sized drone in the field called the Black Hornet Nano, which features a live video feed but a streamlined helicopter-like body shape and rotor blades. While programs have been in place since the 1980s to develop variations on dragonfly, hummingbird (or larger gliding bird), bee, mosquito or beetle-shaped drones, operational drones are unlikely to resemble the insects imagined to be the ultimate self-propelled reconnaissance device. Nonetheless, research and development of small drones draws design knowledge from the study of insects to improve stability, wing shape and action in the goal of achieving efficient, low-powered flight.[37]

Drones' insect-like qualities contribute a great deal to both the discomfort they provoke and their capabilities. Recent theorizing of the nonhuman in philosophy and media theory can help make sense of the conflation or "transposition" taking place between machine and insect. Brian Massumi finds evidence of the fuzzy borders of human and nonhuman in his account of instinct and the "supernormal" qualities of animals and insects, from "the athletic grace of the pounce of the lynx" to the "the architectural feats of the savanna termite" and "the complex weave of the orb spider's web," pointing to their seemingly programmed, "automatic nature, or instinct."[38] While Massumi is concerned with the connections between human and insect, the link to media is made by Jussi Parikka in his study of the "transposition between insects . . . and media technologies."[39] Parikka describes the "uncanny affect" that connects the instincts and movements of insects, robotic machines and algorithmically controlled devices. The artificial intelligence that underpins the stability, maneuverability and automated flight paths or tracking in contemporary drone technology follows this conceptual trajectory. At these points, the semisubjectivity of the drone camera can be felt separately to the operator's control.

Developments in insect-inspired robotics in the form of drones are irksome because they cede something of the division between the instinctual world of "nature" and the artificial power and control of machines, via the threat posed to human life by military industrialism. However, the disruption of communication processes and information capture embodied by these machines also fascinates and inspires. Outside of the militarism that has given rise to the use of drones for aerial imaging, surveillance and drone strikes in theaters of war, two areas of tech development have taken place. On one hand, a whole DIY movement has risen around making, flying and racing drones throughout developed countries. On the other hand, recent commercial developments have revolved around extending the capabilities of drones for augmenting mapping and *inspection*, aiding various industries with the software-enhanced generation of visual data.

How Does a Drone See? Machinic Vision and Semiautomation

Drones *see* through a complicated array of cameras, other sensors and machine vision software systems. The DJI Phantom 4, for example, incorporates a high definition movable camera for recording and real-time signal relay to provide first person view (FPV) control, in addition to two housing-fixed forward-facing stereo vision cameras for automated collision avoidance. In combination with computer vision, object recognition and machine learning processing, these are used to track objects in 3D space, detect and avoid obstacles and track and follow chosen subjects. The sensing system at play in this device not only allows the drone to fly with increased stability but also introduces a semiautonomy that produces new tactical functionality in the machine's ability to respond to visual cues. In order to operate in complex environments, drones need a degree of autonomous visual and sensory processing. And as drone vision improves, the capacity and applications of autonomous flight widen. In this final part of the chapter, we account for the conceptual basis of drones' flight and vision "autonomy," illustrate the essential human link through augmented FPV and point to drones' capacities as platforms for a new modality of geomedia in their visual data capture and analytical functions.

In the 1950s and 1960s, John McCarthy, the computer scientist who coined the term "artificial intelligence," theorized whether it would be possible to program a computer or machine to *learn* about its surroundings rather than being fed preprogrammed information, representations and instructions. He noted in a seminal 1969 paper written with Patrick Hayes that "a computer program capable of acting intelligently in the world must have a general representation of the world in terms of which its inputs are interpreted."[40] It is through commercial drones' return to home (RTH) or autoreturn and precision landing capabilities and their obstacle avoidance technology that they have achieved some of what McCarthy and others envisioned. Return to home systems address the problems earlier models had of becoming untethered once a wireless signal was lost, and so continuing on their flight path never to be seen again. At the time, lost and found forums were in heavy use. RTH uses GPS coordinates to set a flight path back to the controller. DJI built early features to help automate RTH by adding vision positioning, landing protection – which scans the ground to check its suitability for landing – and obstacle avoidance scans (Figure 5.1).[41] By 2018, drones with obstacle avoidance sensors

FIGURE 5.1   DJI Phantom 4, machine vision for automated obstacle avoidance.

Source: DJI.

had become common, with a range of systems used, including lidar, time-of-flight, infrared, ultrasonic, stereo and monocular vision systems. Many integrate data from a number of sensors acting simultaneously (sensor fusion) to compute a "picture" of the drone's surroundings more effectively than any one sensor or data source alone.

McCarthy's contribution to the research field of artificial intelligence was extensive, but from these early days he insisted on bringing together the requisite mathematics and new programming language with philosophy – formal logic in particular – to address underlying assumptions about "what knowledge is and how it is obtained."[42] Since then, others have pre-empted the knowledge capabilities of "seeing machines," most notably, perhaps, in the writings of cultural theorist and aesthetic philosopher Paul Virilio. Virilio provided a critical pre-civilian drone account of machine vision, and it is tempting to take his ideas as a model for understanding contemporary drone vision. His work foregrounds the visual as "a critical site of theory and contemporary cultural action and intervention."[43] Virilio's entry points to this politics of the visual are oblique and could easily incorporate an account of drones as seeing machines. Who or what "sees" and what it means to be seen are certainly central to the politics of drone use and control; and their fuzzy indetermination is previewed by Virilio through many other spheres of contemporary techno-sociality.

Popular and regulatory narratives for drones often imply that their threat lies in what Virilio calls the "industrialization of vision," and in the "splitting of viewpoint," or "the sharing of perception of the environment between the animate (the living subject) and the inanimate (the object, the seeing machine)."[44] Virilio takes his cue from artist Paul Klee: "Now objects perceive me."[45] For Virilio, even in the mid-1990s, "this rather startling assertion has recently become objective fact, truth."[46] Automation of perception hence brings about an "optical imagery with no apparent base, no permanency beyond that of mental or instrumental visual memory."[47] This unnatural splitting of viewpoint produces a new marketplace of the visible. Within Virilio's techno-dystopian scenario, instrumental virtual images are like the foreigner's mental pictures: exclusive and inaccessible. This is the fear of the drone as mechanized vision machine hovering above urban spaces, surveilling and thus affecting relations of public visibility and private seclusion.

But Virilio's error, as John Johnston points out,[48] is to oppose human and technical vision rather than positing visuality, or visualization, as running through and between them. This is what the technical image and the development of machine vision and machine learning brings into play. Drawing from Deleuze and Simondon, Johnston argues that

> in the current climate of accelerated technological innovation, "a new consciousness of the sense of technical objects" may be necessary if we

are to be fully receptive to and engage critically with the new forms and singularities of contemporary visual experience.[49]

Anna Munster also argues that, with technologies like drones, "signal multiplies, yet its relays do not entirely replace the human, rather it passes through and around us, integrating us into its circuits while not relying on us."[50] Hence, with drones we can also ask: what are the new kinds of perception and action or control made possible by our human-technical assemblages? It is the question of the experience of these perceptual systems, or camera consciousness, that is most important to understanding the impact that drone vision has on the scene.

Ultimately, machinic vision should be understood not as a simple matter of seeing with machines, or being "seen" by them – though in a sense this is presupposed and generally assumed – but rather as "a decoded seeing, a becoming of perception in relation to machines that necessarily also involves a recoding."[51] In other words, drone vision is its own special thing and generates new relationships and new forms of visibility and hence altered social configurations. Drone vision incorporates a striving to surmount the eye's immobility, a goal that is at the same time captured within a conceptual bind in which social relations and perceptual capacities are both augmented and radically restructured. We gain a new kind of perception image – the image *of* the perceiving eye-camera – through its foregrounded "dislocation" and aerial motility. To fly drones in first person view, or even to watch the countless of these videos posted to YouTube by enthusiasts, illustrates the elasticity of what we know of the mobility of machine-human vision. And endless drone (and specifically racing drones) forum posts are dedicated to getting the signal, image, control setup for different drones just right.

Drone imaging systems and camera setups vary greatly. Large- and small-scale models make use of video recording and streaming relayed to a ground control location and monitor system via wireless radio frequency transmitters, in combination with satellite GPS tracking and location information. The technological shift toward autonomous seeing or environmental awareness is driven by new visual processing chips. The DJI Mavic, for example, uses the groundbreaking Movidius Myriad 2 chip, which allows for high-powered visual processing at low energy input. The "return to home" function and object or subject tracking this chip enables introduces a new kind of intelligence into drones and gives the camera a new level of visual reflexivity. GoPro cameras are commonly used to transmit high definition video images, often in wide-angle and taking a kind of spherical, global image of ground activity below. This movable camera functions both as one of the primary purposes of drone operations in its aerial media production *and* as its mode of remote visual control through first person view (FPV). Signal strength varies depending on the hardware and network systems but, for civilian drone systems, usually ranges from around 1 to 6 km. In fact, antennae setups account for a great deal of discussion in online drone forums, as operators attempt to build the best kind of signal and vision link for the different circumstances and contexts for their drone flight. For indoor flight, a powerful antennae and receiver can introduce signal echoes that distort or interrupt the FPV. A narrow antenna offers directed slices of signal, a round one disperses the signal over a wider area.

As with the early stages of many new technological systems, a great deal of DIY and community activity across countless forum pages go into adapting and evolving the capabilities of drone gear. YouTube videos abound for testing or showing off that gear. The extensive community forums, YouTube channels, videos testing configurations of gear and FPV or showing new angles on the world through drone vision flights constitutes what Stilgoe refers to as social learning.[52] Social learning or public pedagogy poses new technological configurations as an experiment, revealing exploits, exploring capabilities and sharing knowledge and experiences. First person view equipment and experiences are popular targets of discussion and video sharing. As one forum member puts it, "a great experience depends on control and vision."[53] On Reddit's r/fpv racing subreddit, for example, one member responds to another's questions about FPV equipment with a detailed outline of how to get started. Using an Inductrix FPV + RTF (ready to fly) drone both inside and outside, Ninjajimmy83 suggests, "i'd first practice a bit with LOS (line of sight)," and even before this, "pickup a simulator. I'm using Liftoff on PC (Steam). I found it very helpful and in only a few hours I saw a noticeable improvement."[54] Others discuss problems causing FPV lag or glitches, and how to optimize signal to achieve the best FPV experience.

The popular genre of "FPV freestyle" videos on YouTube is a space for sharing configurations of equipment and drone experience. Videos show often spectacular scenery as it is traversed by rapid drone movement and acrobatics. These are not only about where the drone flies and what it sees, but how. They convey the tumultuous, often spinning or suddenly inverted vision, following hedges or crops in fields, cliffs along a coastline, through trees or buildings, including within their tight interior spaces. In one clip, the drone speeds through an abandoned building, up stairwells, through hallways and rooms in what looks like an old industrial park in Germany.[55] The drone maneuvers around and to the top of a tall water tank tower, "inspecting" its surface, landing, highlighting a small sticker, almost suggesting the key use to which small commercial drones are now being put in their industrial applications.

Aside from the notorious plans companies like Amazon have for using drones to deliver packages,[56] a burgeoning set of industrial and commercial applications are exploiting the imaging, mapping, inspecting and traversing functions of drones. They deliver drone vision in inhospitable and inhuman environments as a product, or simply extend visual analysis capabilities previously carried out by technicians. For example, in construction, vertical scanning augments 2D layered floorplans to feed data into more precise 3D structural models.[57] The Swiss company Flyability has made its mark by developing drones that are better able to carry out inspection and data gathering work in indoor, inaccessible and confined spaces. Its maneuverability technology aids the still essential link between device and operator, demonstrating through its promotional material the benefits of being able to incorporate these direct visual data into reporting and analysis.[58] The company also notes its inspiration is drawn from "the ability of insects to keep their stability after an in-flight collision," while "flight data, thermal video and selected Points Of Interest (POI) are recorded" for later use.[59] Perhaps most spectacularly, Parisian firefighters used DJI Mavic Pro and Matrice M210 drones to track the progression and help fight the fire that engulfed the Notre Dame Cathedral in April 2019.

These industrial applications are supported by an expanding field of research in computer science and robotics that targets drone imaging and vision capabilities. Research papers test the uses of aerial gamma-ray imaging,[60] low-cost mini-UAV thermal and multispectral imaging or radar imaging,[61] for example. Or drones are tested in their use of optical imaging as a tool for spatial assessment, such as in calculating the tree height and canopy crown of forests.[62] Gamma-ray and other forms of radiation imaging systems have been developed and tested for use in inspecting the damaged Fukushima nuclear reactors, enabling visual access to spaces inhospitable to humans.[63] Each of these developments changes the relationship between emergency worker, or researcher, and their subject matter. Drone vision intervenes at the point of generating a new actionable kind of visibility or visual knowledge. As a way of thinking about the redistribution of agency in the age of seeing machines and automated technology, the concept of camera consciousness points to the underlying development that machinic vision introduces. Where FPV racing pushes the limits of seeing with and through hyper-mobile cameras, in these industrial applications, the camera's intelligence combines with their ability to transform the world as visual data.

Conclusion

The work of William H. Whyte, with which we opened this chapter, remains significant for its pioneering use of aerially situated cameras for "scientific" monitoring, assessment and knowledge production. Whyte's use of cameras in this way resonates with the use of drones in cities (and elsewhere) around the world. Drones, as we have outlined in this chapter, have been put to use in order to track and analyze – and, importantly, serve as recorded visual witness to – rallies, congregations, protests and crowd violence among other activities within the city. They have also been employed across various fields and in a variety of capacities for monitoring, inspecting and assessing, as a vehicle for the production of knowledge.

In his influential book *The Practice of Everyday Life*, Michel de Certeau is skeptical of any approach to understanding urban life that pits vision from above, which seeks to render "the complexity of the city readable,"[64] against the lived messiness of life on the street. De Certeau characterizes the former approach, which he refers to as the "solar" or "celestial eye"[65] – as an "atopia-utopia of optical knowledge,"[66] and argues that it fails to capture the complications of the everyday. "Escaping the imaginary totalizations produced by the eye," he writes, "the everyday has a certain strangeness that does not surface, or whose surface is only its upper limit, outlining itself against the visible."[67] At first glance, drones – as motile, camera-equipped, aerial technologies – would appear to fall clearly within the former category – that is, as devices that are firmly within the thrall of the exaltation of the scopic drive and in the service of operators "looking down like a god."[68]

And yet, as we have argued in this chapter, the scopic impacts of drones (especially non-military drones) vis-à-vis occupants of the city is not so straight-forward. Within urban protest situations, for instance, even though the drone camera inhabits aerial space, it does so while acting as an unbounded proxy for those contesting the streets below, and, as such, affects the visual contours of the protesting public. More broadly, the motility of drones complicates relations between object and subject, image and vision, seeing and being seen. As camera-equipped, aerial, motile devices, drones also remain unsettled in their social functionality. First person view (FTV) tethers drones to a dislocated center of control and action, and increasingly guides decision-making from afar – as in their use in warfare, but also for inspection, terrain mapping, agricultural management and forestry. The core argument of this chapter has been that drones and drone use complicate the visual field and lead to a powerful reconfiguration of the territory of the visible, thereby intensifying awareness of the role of the camera in producing new modes of public organization, social life and knowledge. To reiterate a key point, the question of the experience of these perceptual systems, the camera consciousness they affect, remains vitally important to understanding the present and future impacts of drone vision.

Notes

1. Adam Rothstein, *Drone* (New York: Bloomsbury Academic, 2015), 75.
2. William H. Whyte, *The Social Life of Small Urban Spaces* (Washington, DC: The Conservation Foundation, 1980).
3. Two versions of the video appear at: "STUNNING IMAGES: Drone Films Poland Protest Pictures," YouTube video, 1:38. "On Demand News," November 16, 2011. https://youtu.be/gPLh8vkM7ms; and, "Civilian Drone Operated at Polish Riots 11–11–2011," YouTube video, 2:40. "disinpho," November 18, 2011. https://youtu.be/KOxh9dbkNT4.
4. See: "Police Shoot Down RC Quadrocopter in Turkey – Truthloader," YouTube video, 1:45. "Point," June 13, 2013. www.youtube.com/watch?v=_A-ufp5gY3s&t=6s.
5. Ibid.
6. For example: "Aerial Drone Captures Scale of Hong Kong Protests," YouTube video, 0:43. "The Wall Street Journal," September 29, 2014. www.youtube.com/watch?v=0HoEj1BOOpQ.

7. For example: "Bangkok 'Shutdown' 13 Jan 14 Pathumwan," YouTube video, 0:26. "TheCyberJom," January 14, 2014, https://youtu.be/6WWfnRnzZGE?list= PL0PjXgwbk-vur0wGByQSVu9r1ODn5TK8m.

8. See also: Anthony McCosker, "Drone Vision, Zones of Protest, and the New Camera Consciousness," *Media Fields* 9 (2015); Anthony McCosker, "Drone Media: Unruly Systems, Radical Empiricism and Camera Consciousness," *Culture Machine* 16 (2015).

9. Carolyn Marvin, "Your Smart Phones Are Hot Pockets to Us: Context Collapse in a Mobilized Age," *Mobile Media & Communication* 1, no. 1 (2013): 155.

10. Lisa Parks, "Mapping Orbit: Toward a Vertical Public Sphere," in *Public Space, Media Space*, Chris Berry, Janet Harbord, and Rachel Moore (Eds.) (Houndmills, Basingstoke: Palgrave Macmillan, 2013), 61–87; Lisa Parks, *Cultures in Orbit: Satellites and the Televisual* (Durham, NC: Duke University Press, 2005); Lisa Parks, "Vertical Mediation and the U.S. Drone War in the Horn of Africa", in *Life in the Age of Drone Warfare*, Lisa Parks and Karen Caplan (Eds.) (Durham: Duke University Press 2017), 134–157.

11. Parks, "Mapping Orbit."

12. Kevin D. Haggerty and Richard V. Ericson, "The Surveillant Assemblage," *The British Journal of Sociology* 51, no. 4 (2000): 605–622.

13. John Urry, *Mobilities* (Cambridge: Polity, 2007); Gerard Goggin, *Cell Phone Culture: Mobile Technology in Everyday Life* (New York: Routledge, 2006); Nanna Verhoeff, *Mobile Screens: The Visual Regime of Navigation* (Amsterdam: Amsterdam University Press, 2012).

14. The FAA's Small UAS Rule (Part 107) has been operational from 29 August 2016. See: "Unmanned Aircraft Systems (UAS)," *Federal Aviation Administration,* www.faa.gov/uas/.

15. William James, *Essays in Radical Empiricism* (New York: Longmans Green and Co., 1912), 35.

16. Ibid.

17. BBC News, "Gatwick Airport: Drones Ground Flights," *BBC News,* December 20, 2018, www.bbc.com/news/uk-england-sussex-46623754.

18. Parks, *Cultures in Orbit,* 7.

19. Parks, "Mapping Orbit."

20. Cited in Parks, "Mapping Orbit," 65. See: Siegfried Wiessner, "The Public Order of the Geostationary Orbit: Blueprints for the Future," *Yale Journal of World Public Order* 9 (1983): 217–274.

21. Rothstein, *Drone,* ix.

22. Ibid., 75.

23. Grégoire Chamayou, *Drone Theory* (London: Penguin, 2015); Lisa Parks and Karen Caplan, eds., *Life in the Age of Drone Warfare* (Durham: Duke University Press, 2017).

24. See: "Spridningstillstånd," *Lantmäteriet,* www.lantmateriet.se/sv/Om-Lantmateriet/Rattsinformation/spridningstillstand/.

25. Maximilian Jablonowski, "Would You Mind My Drone Taking a Picture of Us?" *Photomediations Machine,* September 29, 2014, http://photomediationsmachine.net/2014/09/29/would-you-mind-my-drone-taking-a-picture-of-us/.

26. Rob Horning, "Selfies Without the Self," *The New Inquiry,* November 23, 2014, http://thenewinquiry.com/blogs/marginal-utility/selfies-without-the-self/.

27. Donna Haraway, *Simians, Cyborgs and Women: The Reinvention of Nature* (New York: Routledge, 1991); Mark Andrejevic, "Becoming Drones: Smartphone Probes and Distributed Sensing," in *Locative Media*, Rowan Wilken and Gerard Goggin (Eds.) (London: Routledge, 2015), 193–207.

28. Adrian Mackenzie, *Wirelessness: Radical Empiricism in Network Cultures* (Cambridge, MA: MIT Press, 2010); Jussi Parikka, *Digital Contagions: A Media Archaeology of Computer Viruses* (New York: Peter Lang, 2007); Anna Munster, *An Aesthesia of Networks: Conjunctive Experience in Art and Technology* (Cambridge, MA: MIT Press, 2013).

29. Parks, "Vertical Mediation," 145.
30. Paul Virilio, *The Vision Machine*, trans. Julie Rose (London and Bloomington, IN: BFI / Indiana University Press, 1994).
31. Ibid.
32. Katherine Chandler, "American Kamikaze: Television-Guided Assault Drones in World War II", in *Life in the Age of Drone Warfare*, Lisa Parks and Karen Caplan (Eds.) (Durham: Duke University Press, 2017), 89–111.
33. Ibid., 91, 94.
34. Ibid., 107.
35. Jay Peters, "Watch DARPA Test Out a Swarm of Drones", *The Verge*, August 9, 2019, www.theverge.com/2019/8/9/20799148/darpa-drones-robots-swarm-military-test.
36. *Eye in the Sky*, directed by Gavin Hood (London: Entertainment One/Raindog Films, 2015).
37. Dario Floreano and Robert J. Wood, "Science, Technology and the Future of Small Autonomous Drones," *Nature* 521, no. 7553 (2015): 460.
38. Brian Massumi, "The Supernormal Animal," in *The Nonhuman Turn*, Richard Grusin (Ed.) (Minneapolis, MN: University of Minnesota Press, 2015), 1.
39. Jussi Parikka, *Insect Media* (Minneapolis, MN: University of Minnesota Press, 2010), xiii.
40. John McCarthy and Patrick J. Hayes, "Some Philosophical Problems from the Standpoint of Artificial Intelligence," in *Readings in Artificial Intelligence*, Bonnie Lynn Webber and Nils J. Nilsson (Eds.) (Burlington, MA: Morgan Kaufman, 1981), 431.
41. See, for example: DroneZon, "12 Top Collision Avoidance Drones and Obstacle Detection Explained," *DroneZon*, December 28, 2018, www.dronezon.com/learn-about-drones-quadcopters/top-drones-with-obstacle-detection-collision-avoidance-sensors-explained/.
42. Ibid., 431.
43. John Armitage and Ryan Bishop, eds., *Virilio and Visual Culture* (Edinburgh: Edinburgh University Press, 2013), 1.
44. Virilio, *The Vision Machine*, 59–60.
45. Ibid.
46. Ibid.
47. Ibid.
48. John Johnston, "Machinic Vision," *Critical Inquiry* 26 no. 1 (1999): 27–48.
49. Ibid., 27; Gilbert Simondon, *Du Mode d'existence des Objets Techniques* (Paris: Aubier, 1958).
50. Anna Munster, "Transmateriality: Toward an Energetics of Signal in Contemporary Mediatic Assemblages," *Cultural Studies Review* 20 no. 1 (2014): 150–167.
51. Johnston, "Machinic Vision," 23.
52. Jack Stilgoe, "Machine Learning, Social Learning and the Governance of Self-driving Cars," *Social Studies of Science* 48, no. 1 (2018): 25–56. For a discussion of the use of drones as visual method, see: Bradley L. Garrett and Anthony McCosker, "Non-Human Sensing: New Methodologies for the Drone Assemblage," in *Refiguring Techniques in Digital Visual Research*, Edgar Gómez Cruz, Shanti Sumartojo, and Sarah Pink (Eds.) (Basingstoke: Palgrave Macmillan, 2017), 13–25.
53. Futuramille, "First Person View Equipment," *Drone Racing Pilots Forums*, September 8, 2018, https://droneracingpilots.com/threads/kinda-bummed-flysight-spexman-2-goggles.2064/#post-18942.
54. Ninjajimmy83, "Getting Started in FPV – Advice or Recommendations," *Reddit, r/fpvracing*, April 22, 2019, www.reddit.com/r/fpvracing/comments/bfsgwb/getting_started_in_fpv_advice_or_recommendations/.
55. "Why Should You Fly Freestyle at 800mW? | FPV," YouTube video, 0:15. "Mr Steele," July 28, 2018. www.youtube.com/watch?v=bBb_kSO3vTo&t=105s.

56. Jaihyun Lee, "Optimization of a Modular Drone Delivery System," in *Proceedings of 2017 Annual IEEE International Systems Conference (SysCon)* (New York: IEEE, 2017), 1–8. See also: "First Prime Air Delivery," *Amazon Prime Air*, www.amazon.com/Amazon-Prime-Air/b?ie=UTF8&node=8037720011.

57. See, for example: Geospatial World, "Automatic Vertical Scanning for Drones Now Available," *Geospatial World*, February 22, 2019, www.geospatialworld.net/news/%D0%B0utomatic-vertical-scanning-for-drones-now-available/.

58. Flyability, "Elios – Inspect & Explore Indoor and Confined Spaces," *Flyability*, www.flyability.com/elios/.

59. Ibid.

60. Yuki Sato, et al., "Remote Radiation Imaging System Using a Compact Gamma-ray Imager Mounted on a Multicopter Drone," *Journal of Nuclear Science and Technology* 55, no. 1 (2018): 90–96.

61. Juliane Bendig, Andreas Bolten, and Georg Bareth, "Introducing a Low-cost Mini-UAV for Thermal- and Multispectral-Imaging," *International Archives of the Photogrammetry, Remote Sensing and Spatial Information Sciences* 39, no. B1 (2012): 345–349.

62. Ye Seul Lim, et al., "Calculation of Tree Height and Canopy Crown from Drone Images Using Segmentation," *Journal of the Korean Society of Surveying, Geodesy, Photogrammetry and Cartography* 33, no. 6 (2015): 605–613.

63. Saku Mochizuki, et al., "First Demonstration of Aerial Gamma-ray Imaging Using Drone for Prompt Radiation Survey in Fukushima," *Journal of Instrumentation* 12, November (2017), https://doi.org/10.1088/1748-0221/12/11/P11014.

64. Michel de Certeau, *The Practice of Everyday Life*, trans. Steven Rendall (Berkeley, CA: University of California Press, 1984), 92.

65. Ibid.

66. Ibid., 93.

67. Ibid.

68. Ibid., 92.

6

HOW DOES A CAR LEARN TO SEE?

In 2017, a number of press reports made light of the fact that while European car manufacturers designing autonomous vehicles seemed to have solved the "moose test" (detecting moose and deer as they cross roads and adequately predicting their behavior so as to avoid hitting them), they had great difficulty passing the "kangaroo test."[1] This was due to a lack of training data and a kangaroo's more erratic jumping movements. In the industry these are called corner cases – the rare, unpredictable actions of animate or inanimate objects that move in complicated ways. In Australia, one car insurance company reported close to 8,000 accident claims involving kangaroos between 2018 and 2019.[2] Despite the overall numbers, these kinds of incidents are understood as limit cases for systems that have to see and respond with awareness constructed from prediction, inference and learned knowledge. The consequences of what an autonomous vehicle (also referred to popularly as self-driving cars; hereafter shortened to AVs) can and cannot see are serious. But the more pertinent and vexing issue is *how* an AV sees, or rather how it *learns* to see and act, and whether these processes can ultimately improve driver safety and affect better transport systems.

In this chapter, we explore the capabilities – the ambitions and realities – of AVs against some of the key challenges they face in entering urban environments and everyday use. Learned responsiveness of the sort that will allow an AV to adjust to a sudden unpredictable situation is symbolic of the as yet unattainable pinnacle of fully competent artificially intelligent machine vision systems. It is no surprise then that there is little agreement about the lead time for fully autonomous vehicles to be operational, let alone agreement about their impact on the urban environment and on society. And while the capabilities of machine vision, and the deep learning systems it is built upon, are nowhere more heavily scrutinized than in their use for guiding AVs, they remain opaque, often obscured by commercial competition.

Significant attention has been paid to the safety concerns and failures associated with removing human control from moving vehicles. However, fundamental to the long-term success of AVs is the development of visual capture and processing – that is, the procedures that are involved in the mapping, sensing and real-time visual data processing within dynamic urban environments. This is where machine learning comes into play, but what the case of AVs shows us about the quest for automated vision is that camera consciousness (the interaction of vision–awareness–action in specific contexts) is all about the relationship between bodies, and their relationship to an environment. To be successful, vehicles, their computational control systems, physical environments and infrastructure and people have to learn together. Vehicles described as autonomous or self-driving are often heavily "connected." They are certainly beginning to reduce the need for in-car hands-on human control but remain intimately connected with an urban infrastructure of embedded sensors and transmitters, coordinated dataflows and signaling, all working together through vehicle-to-vehicle and vehicle-to-infrastructure multiway communication.[3]

Examining vision capture and processing by AVs, we apply James Gibson's ecological theory of perception to ask *how* a driverless car learns to see. In addition, we explore how visual processing technology embeds human and social problems in the path of developers. Vision capture and machine vision systems have to negotiate roadworks and "unexpected" obstacles such as moose and kangaroos. But they also have to negotiate human behavioral challenges – such as predicting and understanding other road users, for instance, and reacting to and appropriately negotiating all the subtle *cultures* of driving, the often subtle movements and indications that are the cultural codes through which cars, people, bike riders, pedestrians and so on act. These challenges are increasingly entwined with an ecosystem of public or proprietary data capture, processing and access, as tech companies and subsidiaries (like Waymo, Alphabet's autonomous vehicle arm) jostle to claim stakes in the infrastructure that will make connected AVs a reality.

Our argument is that these are, in an expansive sense, ecological problems of the functioning of perception, sense and meaning-making. Automated vision, understood through this lens, is ecological, relational (a set of dynamic, topological relationships) and contextual, embedded in situations that constantly change what machine vision is and sees. We begin by exploring the shifting groundwork around which the connected data ecosystem of AVs is currently being colonized and commercialized. Second, we look at some of the myths of autonomy that emerge out of public spectacles like the DARPA Grand Challenge events, in which teams competed in races to test and show off their vehicle's capacity to successfully traverse real-world spaces. Finally, we illustrate some of the entrenched challenges AVs face as they disrupt the network of relationships that vehicles, their occupants and transportation infrastructures continuously and dynamically generate.

The key point often lost on those heralding the potential of AVs for reshaping our urban environments is that a truly autonomous vehicle requires a kind of shared camera consciousness that reconfigures the perceptual ecosystem well beyond the technology embedded in the vehicle itself – the cameras, lidar and sensor systems – to involve a whole digital communication infrastructure and the social context of transportation. AVs are already an assemblage of connected technologies (vehicle-to-vehicle and vehicle-to-infrastructure), and equally importantly but often underestimated, of drivers, passengers, other road users, animals and other obstacle bodies and weather conditions. The race to develop some successful form of extended or full automation in vehicle control systems is underpinned by a delicate politics of machine vision within a disrupted social semiotic assemblage.

Mapping the Visible Terrain: Seeing, Sensing and Perceiving

According to a *SlashGear* article in late 2015, three key questions must be answered at all times and in real time for a self-driving car to do what it does: "where the car is, what's around the car, and what should the car do next based on the first two questions."[4] These actions might be further summarized as orientation, perception and decision. The US National Highway Traffic Safety Administration (NHTSA) has its own definitions of automation in its Automated Vehicle Policy, where it defines six levels of automation.[5] In the NHTSA scheme, level zero is attributed to vehicles with no automation. Levels one and two refer to driver assistance components for either steering or acceleration/deceleration using information about the driving environment but with human driver performance at all time. Level three involves "conditional automation," with a human driver responding only to an alert to intervene. Level four, or "high automation," places the vehicle in control of "all aspects of the dynamic driving task, even if a human driver does not respond appropriately to a request to intervene."[6] And "full automation," level five, incorporates full vehicle control "under all roadway and environmental conditions that can be managed by a human driver."[7]

As a schematic, the NHTSA model of the levels of autonomy involves shifts in capacity for seeing and acting in an environment: monitoring → (environmental conditions) → control (machine, human or shared) → performance (vehicle-centered, risk and safety modulations). Our initial point is that like humans, an autonomous vehicle does not see, or at least see effectively, outside of environmental, social, cultural, spatial, historical (learned), infrastructural and otherwise mediated contexts. This is why the terrain of visual and spatial data is expanding and holds great value for technology companies and vehicle manufacturers in a rapid phase of colonization on the way to what is currently understood as level five autonomy.

All AVs contain a variety of tools to perform those three operations (orientation, perception and decision). While the exact arrays of sensors and other technologies mounted onto or integrated into AVs differs from developer to developer, in general terms these systems incorporate some combination of the following: video cameras, radar sensors, ultrasonic sensors, lidar (light detection and ranging) sensors, a GPS unit and an on-board central computer. Video cameras are used to detect traffic signals, read road signage and keep track of other vehicles and record the presence of pedestrians and obstacles. Radar sensors determine the position of nearby vehicles, while ultrasonic sensors, which tend to be wheel-mounted, are used to measure the position of close objects, such as curbs and other cars when parking. Lidar sensors bounce pulses of infrared light off of surrounding objects, and these pulses are used to identify lane markings and road edges. GPS signals are combined with readings from other internal sensors and measurement devices (tachometers, altimeters and gyroscopes) to provide positioning information. All of this data is processed and actioned by the car's on-board computer system, with algorithms and software, linked with cloud data and processing, vital for the decision-making element of AVs.

All of the tools detailed earlier play a vital role in calculating orientation, perception and decision. GPS is obviously a vital component in determining orientation, as are on-board maps. Orientation is also calculated through the use of radar and lidar sensors. However, due to the unreliability and inaccuracies of GPS signals and lidar sensor data, and the need for high-precision orientation information at all times ("an autonomous car that is 99% safe won't be good enough"),[8] AVs supplement these technologies by deploying sophisticated cameras and other direct sensors to navigate, analyze and make decisions in real time about how to proceed. Less obvious in the evolution of AVs is the role played by mapping platforms and data processing. However, no matter how sophisticated the array of perceptual technologies, AVs can only function effectively in an environment that "speaks back," even while they continue to "feed in" to a real-time, dynamic urban map.

In recent years, tech journalists, automotive analysts and YouTube enthusiasts have circulated countless lidar videos to provide a sense of what an autonomous vehicle "sees" as it traverses an environment, orienting itself in relation to the unfolding 3D digital map.[9] As with the communities of drone vision video sharing, these form part of a collective space of public learning[10] as people seek to understand the new perceptual fields AVs are creating (a point we return to in the final section). Although lidar technology has been prohibitively expensive, it offers an important component of an autonomous vehicle's sensing systems, not least because it can overcome some of the problems associated with cameras in distinguishing surfaces and measuring depth perception (see Figure 6.1), problems we address in more detail in the following section. As a technology that adds to the visual map that guides AVs, lidar highlights the digital communication ecosystem that will come to alter the visible terrain of our urban

FIGURE 6.1 Lidar vision.

Source: Luminar Technology.[11]

environments. The interesting development here is that while lidar has been used for some time in aerial and undersea mapping (including drone mapping), its use in AVs is not only for mapping but also for traversing a terrain mapped by informing activators in the steering, acceleration and braking systems.

In addition to these automotive applications of lidar technologies, Uber's purchase in 2015 of mapping start-up deCarta, and the sale the same year of Nokia's HERE mapping assets to a consortium of German car companies, were perhaps early signs of how mapping infrastructures have become highly prized technology commodities for addressing problems of orientation.[12] In early 2017, Intel made its move to join other technology companies in building infrastructure for AVs and Internet of Things systems. It paid US$15.3 billion for Mobileye, a company that makes forward- and rear-facing cameras that capture imagery for Advanced Driver Assistance Services (ADAS) – automated parking, obstacle avoidance and proximity indicators – for more than 25 automotive manufacturers. Intel's interest in the company lay in its move to collect "localization" or positioning data through more than 2 million cars already on the road, as a kind of crowdsourced "Roadbook" or "continuously updating map" as analyst Mark Prioleau puts it.[13] "The Roadbook," Prioleau writes, "contains information about everything relevant about the road: the streets, the lanes, the buildings, signs and features that sit by the roadside and the data about how vehicles typically move through that environment."[14] By investing in this live updating map, Intel is buying into the flipside of the autonomous vehicle as seeing machine: the data processing infrastructure that enables it to act within an actual complex and dynamic environment, extending the notion of the car as dynamic communication platform.[15]

In a PyTorch video in 2019, Tesla's head of AI and computer vision Andrej Karpathy noted that, famously, Tesla doesn't currently use lidar and doesn't use high definition maps "so everything built for Autopilot comes from machine vision and machine learning and the raw video streams that come from the eight cameras that surround the vehicle."[16] Deep learning systems underpin the camera components of AVs as seeing machines. What does it mean to train a machine vision system? One such system, SegNet, developed for AVs at Cambridge University, "takes an image of a street scene it hasn't seen before and classifies it, sorting objects into 12 categories – roads, street signs, pedestrians, buildings, cyclists – in real time. It can also deal with light, shadow, and night-time environments."[17] The system was initially trained by undergraduates "who manually labelled every pixel in 5,000 images," taking two days to educate the system, resulting in a labeling accuracy of 90% of pixels.[18]

As one of many different systems designed to produce actionable meaning from pixel data, SegNet was developed to automate the process of encoding and decoding vision from a moving vehicle (adaptable to more complex environments still such as building interiors).[19] This kind of system, however, is only one potential component of the machine vision system needed for an AV. A classifier like SegNet's has a fixed, limited or human-trained set of classes that it is capable of detecting. What's needed is a system that learns on the go. Researchers from visual processing chip maker NVIDIA Corporation, for instance, sought to develop a system that builds trained datasets through a car's steering action.[20]

These technological developments and corporate acquisitions suggest that AVs can only be understood in terms of the sensing technologies that place them in relation to a dynamic environment, through an ecosystem of relational perception. For psychologist and theorist of perception James Gibson, the perceptual ecosystem is dynamic, relational and deeply pragmatic.[21] Published decades before the coalescence of sensing and processing technologies developed for AVs, Gibson's ecological theory of perception was concerned with how people perceive the world around them not by constructing mental representations from the raw materials of experiences, nor like a data processing device. Rather, for Gibson, perception is the outcome of the body's movement through and potential interactions with its environment, and a reciprocal interplay of the two systems of objects understood relationally, in terms of what they "afford" each other. It is the action-potential established in the relationships between bodies (animate or inanimate) in their environment.

The ecological vision needed for an AV to operate in an unpredictable environment can, again, be illustrated by the "kangaroo problem." As Volvo Australia's managing director David Pickett points out:

> When it's in the air, it actually looks like it's further away, then it lands and it looks closer. . . . Because the cars use the ground as a reference

point, they become confused by a hopping kangaroo, unable to determine how far away it is.[22]

Importantly, affordances like the width or curve of a road or the qualities of
its surface only frame rather than determine the possibilities for its use.[23] An
"affordances" approach to perception and the perceptual environment empha-
sizes the relationship a body has with the "data" it receives from the world. In
this sense, the whole environment that an autonomous vehicle operates in can
be understood as a medium that enables, constrains and communicates safe or
unsafe movement. As a system of sensors, feedback mechanisms, movement
and information processing, the autonomous vehicle's primary role is to continuously identify and negotiate what Gibson calls the "margin of safety," the
distance between itself "and the brink."[24]

Gibson's understanding of perception and affordance is not necessarily
orthodox despite its influence in fields such as design and technology studies. AVs, by necessity, have to put into operation many of his central assertions
about perception and environmental affordances. And in part the fascination
with what an autonomous vehicle can see is a response to the imagined subjective capacities of a machine ordinarily conceived as an object (or even at times
an extension of the body of the driving subject).[25] For Gibson, an ecological
understanding of perception dispenses altogether with a subject–object-centric
account of the world for one that is relational. That is, things in the world do
not possess essential, objective properties but rather affordances that arise in the
fluid relations between objects, surfaces, animals: "an affordance cuts across the
dichotomy of subjective-objective. . . . [I]t is equally a fact of the environment
and a fact of behavior."[26] This affects the way we see an autonomous vehicle,
and the way we can understand how it sees the world or what it sees.

If we take a pragmatic approach to how an AV sees, it's more accurate to
acknowledge that it does not see objects, signs, people as such. It sees and must
continuously process the relationship between itself, its positioning data and
drive components and certain attributes or affordances of each of those things
it encounters. As a meaning-making, or semiotic system, these relationships are
established pragmatically in relation to the manner of its seeing and the actions
they afford the vehicle. A car doesn't need to see "ice" on the road but rather
compute the variations in surface traction that alter its capacity to stop or steer.
If a stop sign has been defaced or its surface bent, it may no longer afford the
vehicle a signal point to activate breaking to zero movement, or it may make no
difference to the vehicle's capacity to make that decision and act.

Seeing Vehicles and Urban Obstacles

Fundamental to the success of driverless vehicles, as we have been arguing
throughout this book, is the machine vision that allows real-time, human-like

recognition of environmental cues, informing decision-action systems for machine control. The challenge, of course, is the computational power and precision needed for a machine to adjust and adapt to changing and often unpredictable urban environments. Not only this, but as noted at the outset of this chapter, autonomous machines have to adjust to the actions of humans and other animals. These issues also fuse technological problems with human and social problems. Complex environmental challenges set AVs apart, for instance, from passenger airlines, which have deployed "autopilot" sequences and flight paths for many years now, or automated mining trucks used within the controlled environment of some opencut mines.

An important step, then, has been to better understand those complex environments, conceiving them as a set of problems to be solved. One of the "origin stories" for the revived hopes of developing AVs was the DARPA (Defense Advanced Research Projects Agency) Grand Challenge events of the early 2000s. These challenges connected the competitiveness of start-up culture with the innumerable technical and other hurdles that had to be overcome in order to create workable AVs. The US DARPA-sponsored Grand Challenge events were run three times, offering generous cash prizes, designed to attract talent, and hasten the development of autonomous vehicle technologies. While they did not result in fully competent AVs, they did leave a trail of evidence for the interaction between the vehicles and the natural environment, and the litany of obstacles and hazards that were to run so many teams off the road.

In the first Grand Challenge, staged in the Mojave Desert in 2004, no entrant managed to complete the course. In the second, staged in 2005, participating vehicles were required to negotiate a winding narrow pass, three narrow tunnels and some switchbacks; five entrants managed to successfully complete the course. The third event, known as the "Urban Challenge," was held in 2007 at an old air force base. Participating vehicles had to demonstrate that they could operate in traffic, obeying road rules, while negotiating other vehicles and various obstacles. This event had clear first, second and third place-getters. None of the three events passed without any incident. In the first, one vehicle disappeared into the sagebrush "seven and a half miles into the 150-mile course," and subsequently "burst into flames"[27]; in the second, a vehicle eventually completed the event after overcoming a "pause" to the autonomous system that left it idling on the course all night[28]; while in the third, one vehicle struck a building while another had its Velodyne lidar sensors "decapitated" when it failed to see a fold-over boom gate.[29]

The published academic journal "field reports," produced by entrants in the DARPA Grand Challenges, make for fascinating reading, providing technical insight into just how demanding the challenges were in building "intelligent software" for successful AV operation in complicated urban environments. For instance, the team responsible for one entrant in the Urban Challenge event gives a detailed account of the difficulties associated with creating systems to determine road lane positioning and lane "splines" or curvature ("the splining process always provided smooth paths for the vehicle but lacked the realization of the actual geometry of the roads and intersections").[30] A further field report relays an incident where there was confusion between two autonomous cars to illustrate just "how difficult it can be for a robot [AV] to distinguish between a car stopped at a stop sign and a car parked on the side of the road."[31] Yet another discusses the "inefficiency" of "raw sensor data," which can have a low signal-to-noise ratio, as much lidar data come from "ground readings, curbs and tree tops."[32]

What is especially striking about these field reports in the context of this book is the primacy of vision in the design of these autonomous vehicle systems. This is particularly evident in the software architecture planning diagrams provided for "Boss" and "Odin," the vehicles placed first and third, respectively, in the 2007 Urban Challenge event. In both vehicles, "perception" is the primary component that interfaces with the various sensors, "fuses the data into a coherent world model" and communicates this information to the mission or route planning, motion planning and vehicle behavior components of the system.[33]

There are two reasons vision takes primacy among the array of driverless car sensors. The first is that it plays a vital role, not just in calculating vehicle position, but also in locating and identifying obstacles in the vehicle's path. For instance, in the 2005 Grand Challenge, one vehicle, "Prospect Eleven," "relied solely on stereo vision to detect and range obstacles," while other systems in the 2007 event used the infrared vision of lidar for sensing the surrounding environment and determining such things as lane markings.[34] The second reason – and arguably the most important in Gibson's sense – is that the perception components play a crucial ecological role for these vehicles in determining "what changes take place in the environment over time" and in as close as possible to real time.[35] This is to say that visual perception technologies are fundamental to the automotive body's movement through its environment, and part of the reciprocal interplay and relationality between this automotive body and the obstacles and other features of the environment through which it moves.

Subsequent to the DARPA Grand Challenges, one of the urban transport contexts that has proven especially challenging for developers of AV systems is roadworks. Roadworks represent the unpredictable, difficult to code breakdown of the road as traversable space. A deep pothole, a barrier or a sign with unusual markings, can throw an AV off its path. As Aarian Marshall explains, "this gets to the central challenge of autonomous driving: How do you teach machines to deal with the chaotic, grubby humanity of our roads, where the rules bend so easily?"[36] While human drivers can understand and respond to "any improvised arrangement of placards, blinking arrows, and sign-spinning workers," these signals are much more difficult for a computer to interpret.[37]

In addition to the physical and semiotic complexities faced by AVs as they navigate urban space (such as roadworks), cultural subtleties in driver behavior present just as many (if not greater) challenges. One of the known problems with AV systems is in reading the subtle "human" behaviors of other cars on the road. These challenges become particularly acute at key "trigger points" (as engineers working on AVs call them), those moments where the car's computer is required to decide, such as when merging at intersections. Complications come where there are cultural differences shaping driving behaviors,[38] or where there's a tacit knowledge and set of assumptions at play, and subtle signs that a car will hold back to leave a space. It comes as no surprise, then, to hear that the developers of "Little Ben," one of the entrants in the 2007 DARPA Urban Challenge, counted as a major success the fact that their car entered a four-way intersection with other AVs and was able to maneuver around a car (which had stopped temporarily on the wrong side of the road while trying to pass other traffic) and then return to its correct lane:

> Whereas the correctness of this behaviour may seem obvious, what is significant is that, of the four robots that appeared at this intersection, only Little Ben appeared to proceed as a human operator would have done.[39]

What are being described here are questions concerning what it is for a car to "see" *and* the ecological implications of that computer-determined perceptual and decision-making activity. The designers of these systems are working within a set of ecological problems and are responding with clear reference to the natural environment and animal perceptual systems with the aim of actualizing an "insect media" as Parikka theorizes it, and as we saw in the context of drones' swarming and motile movements, or drawing on a wider range of nonhuman perceptual systems and mechanisms to solve the problems of traversing a complex natural environment.[40]

It may not be by chance that many of the research papers published in the wake of the DARPA Grand Challenges consistently draw on biological and ecological language to describe the perceptual and processing mechanisms of AVs. Extending pioneering German research on AVs as seeing machines displaying "saccadic vision" (the simultaneous movement of both eyes as they jump rapidly from one point of focus to another),[41] one presentation talks about "foveated imaging,"[42] a term that relates to fixation points at the center of the eye's retina, or fovea. Another discusses "exteroceptive sensing,"[43] where exteroception relates to stimuli produced outside of an organism and the sensing that relates the organism to its surroundings. In addition to these examples, there is other language that further draws out the agentic qualities of these vision systems, such as discussion of how on-car systems display "selective" or "prioritized attention,"[44] particularly in order to overcome sensor-generated environmental "noise."[45] The language of "trigger points" indicates decision-making in a way that signals an organism-like agency, not simply an autonomy of movement, but a capacity to "see" and "anticipate" objects that are hidden.[46]

The various combinations of mapping and sensing technologies that have been developed to enable autonomous systems have already begun to alter the boundaries between natural media – the surfaces, substances, groundings – and the visible environment as media, as digital data. Rather than dealing with an object in a gridded space, this situation exemplifies Parikka's point, following Gibson, about the "specificity and positionality of any act of perception, and hence the intensity at which perception is intimately connected to motility."[47] An autonomous vehicle must operate through a "double articulation" of the array of perceiving sensors and the data-map, or the urban milieu that constitutes the medium, "both taking part in the commonality of their event, shaping each other, and changing with each other."[48] In this context, roadworks and other strange obstacles present, as Gibson would have it, an ecological problem of the functioning of perception. Most important for the development of AVs' sensory systems is the socialization of the learning process – through events and data and knowledge sharing such as the Grand Challenges, and by maintaining the kind of openness and responsiveness that can be understood as a coevolving camera consciousness.

Collectively Machine Learning

Proponents of AVs point to the fallibility of human drivers, their tendency, for instance, to swerve to avoid animals potentially resulting in more serious accidents for themselves. AVs can, the theory goes, bring a greater level of safety through computational decision-making not tied to pesky moral decisions about whether to hit a kangaroo or swerve off the road. However, it was reports of the first known fatality involving a Tesla self-driving car in 2016 that raised substantial questions about the safety and trustworthiness of AVs, and in particular, their ability to adequately see and respond to the world around them. As Jack Stilgoe (2018) has explained, a significant emphasis was placed on "driver error" and human fallibility in managing the shared control expected of an *autonomous* vehicle on *autopilot* mode. Machine vision, underpinned by deep learning, flips how we normally understand software and computing, from the rule-based frameworks of if/then to rule-finding and pattern recognition processes. Machine vision mostly uses large amounts of categorized data to make sense of new inputs through "brute force," creating the "means of engaging with an uncertain world that is impossible to capture with a set of formal rules."[49] There is, Stilgoe argues, a great opportunity to democratize learning in the development of AVs, to resist "the story of autonomy is a way of downplaying a car's connections with other vehicles, the built environment and the infrastructure of regulation."[50] It is possible to socialize machine learning. More often, however, the relationship between vehicle, people and the environment in which it operates is conceived in opposition and even in tension.

In the 2016 Tesla crash, the vehicle reportedly did not "see" a truck crossing its path, hitting the trailer at 74 mph, killing the driver on impact. Reports and statements from Tesla referred to the inability of the car's vision system to distinguish the white trailer against the skyline behind it. Tesla's official statement about the incident is interesting and troubling for many reasons.[51] It emphasized that Autopilot was still in "a public beta phase" and that "[n]either Autopilot nor the driver noticed the white side of the tractor trailer against a brightly lit sky, so the brake was not applied," sharing responsibility for driving the car equally between Autopilot and the driver. The statement clarified at length what is expected of a driver, that Autopilot "is an assist feature that requires you to keep your hands on the steering wheel at all times," that "you need to maintain control and responsibility for your vehicle" and "be prepared to take over at all times." The system, it goes on to note, "makes frequent checks" to ensure the driver's hands are on the wheel, delivering "visual and audible alerts if hands-on is not detected."

The name "Autopilot" was criticized in investigations for itself signaling a level of automation not commensurate with its beta phase and its subsequent instructions for use. Appended to the warnings in Tesla's statement, however, is a hope and vision for future use, acknowledging the tragedy but imagining a pathway for a form of automated vision or camera consciousness that does not require such human intervention or co-awareness:

> As more real-world miles accumulate and the software logic accounts for increasingly rare events, the probability of injury will keep decreasing. Autopilot is getting better all the time, but it is not perfect and still requires the driver to remain alert. Nonetheless, when used in conjunction with driver oversight, the data is unequivocal that Autopilot reduces driver workload and results in a statistically significant improvement in safety when compared to purely manual driving.[52]

Humans, of course, can have poor vision too, and poor judgment, or can be impaired by alcohol or fatigue, or can simply be distracted. For Stilgoe, "A technology that works well right up to the point that it doesn't, particularly when that point demands the attention of a user who has lost concentration, represents a substantial regulatory problem."[53] Such a system confers a disproportionate capacity for machine–environment awareness and reaction upon the vehicle's driver who has simultaneously been encouraged to reduce that attention and cede control. This scenario illuminates AV designers' inability to conceive an adequately integrated and altered human–machine camera consciousness.

Tesla's 2016 crash statement cited earlier points toward several core components of the problem of the coevolving camera consciousness at stake in the development of AVs. While driverless systems are "getting better all the time," they nonetheless require driver involvement and attention. The pathway to addressing these component problems involves a set of interactions between machine, human, infrastructure in a model of social learning and cooperation. Stilgoe draws on the public communication and regulatory maneuvering around this incident to argue for the role that social learning can and needs to play in democratizing machine learning and managing public involvement in developing autonomous vehicle systems. He cites Bandura (1988) and Dewey (1916) as some of the key sources of a rich and extensive scholarship on social learning, as a way of considering what an effective, accountable machine learning could be:

> Both senses of social learning – how people learn socially and how societies learn – apply. The latter is more obviously relevant to the governance of new technologies. But the former becomes particularly interesting when we consider the "social" within machine learning.[54]

Combining the social learning processes surrounding development of AVs with Gibson's affordance approach to perception as environmental, we can start to see the extent to which AVs are embedded in and are affecting a coevolving camera consciousness.

Again, the infrastructure of communication is essential to the contextual and social learning of AVs. Researchers at the University of Michigan Transportation Research Institute (UMTRI),[55] for instance, place an emphasis on connected vehicles and vehicle-to-vehicle and vehicle-to-infrastructure communication systems, as well as traffic modeling. Combined with other aspects of vehicle safety, and research examining driver behavior and that of pedestrians and cyclists, the approach of UMTRI is to treat AVs as part of a connected communication ecosystem rather than as strictly "autonomous." The surrounding town of Ann Arbor has become one of the most infrastructure rich AV training environments in the United States, dense with sensors for testing and a high-tech traffic signal control system. This is all recognition that the vehicles themselves can't "go it alone." Ironically, extensive urban infrastructure and modeling of human behavior both in and around AVs are crucial to operational level four and five autonomy. Through this research, the unevenness of AV receptive environments becomes more apparent, as do the cultural and human codes that affect and are embedded in the operation of all vehicles, including AVs. AVs can't operate effectively, for instance, if they do not understand the range of behaviors typical of cyclists or other vehicles at crucial points of interaction, such as at intersections.

Taking cues from the surface features of the environment that indicate the path of other bodies offers a means for contextualizing the perceptual systems of AVs. A series of research projects at MIT's Computer Science and Artificial Intelligence Laboratory (CSAIL) are trying to extend the visual capacity of AVs by better understanding the relationship between vehicles and the visual cues surrounding it. One project looks at the use of shadows at the edge of approaching corners to predict impending movement. The ShadowCam system "uses sequences of video frames from a camera targeting a specific area, such as the floor in front of a corner. It detects changes in light intensity over time, from image to image, that may indicate something moving away or coming closer"; "ShadowCam computes that information and classifies each image as containing a stationary object or a dynamic, moving one. If it gets to a dynamic image, it reacts accordingly."[56] Another project addresses the problem of "occluded" intersections, where there are no traffic lights to aid entry, and where an AV has to see the road from several directions, understand the movement of cross-path vehicles, calculate the time available for merging and manage that merge while dealing with the "behaviors" of the vehicles around it. Human interactions also play a prominent role in the social contexts that AVs have to navigate. Estimating the risk of collision in these situations involves weighing a number of factors,

> including all nearby visual obstructions, sensor noise and errors, the speed of other cars, and even the attentiveness of other drivers. Based on the measured risk, the system may advise the car to stop, pull into traffic, or nudge forward to gather more data.[57]

The crucial link between camera awareness and responsiveness, central to our concept of camera consciousness, is troubled by the distribution of dynamic relationships when vehicles interact in lively environments. The learning required to adapt to AVs is shared and social. In this sense, "acceptance" of AVs on the roads among other road users is more than merely a matter of attitudes and opinions but encompasses an active willingness to incorporate something new into an existing context.[58] Likewise, Stilgoe notes the tension surrounding whether or not AVs should be labeled so as to alert other road users, whether those road users might take advantage of AVs or even interact erratically on purpose, as some researchers speculate.[59] This perceived "need for vehicles to actively communicate their presence" tells us something about the entangled and coevolving camera consciousness involved in the integration of AVs into existing transport environments.

Many hours of YouTube videos offer another venue through which social learning regarding AVs, or their "socialization," takes place. Along with fascination over AV crashes, YouTube is saturated by technology and automotive reviews of AVs, and many advocates and enthusiasts who take AV vision and driverless experience videos to YouTube. Stilgoe sees this as "a form of alternative online pedagogy."[60] Like the countless drone vision videos discussed in Chapter 5, self-driving car videos commonly show an in-car view often adorned with the car's sensory display as it translates the visual environment into machine readable objects in relation to which it responds in real time.[61] It is possible to find within these public pedagogies spaces for intervention in machine learning, error identification or near-miss scenarios. Some have used the vast hours of such YouTube videos to study the "social interaction" and performance of AVs.[62]

Conclusion

James Gibson warned against simplistic models of a body's perception of the world around it. The process was too often understood as if it were a matter of constructing a mental image or representation from "raw" sensory data, in a linear information or data processing systems sense. He saw perception as the result of the body's movement and potential interactions, as a reciprocal interplay of multiple bodies' facets, planes and movements in relation to what they *afford* each other. Whether attempting to deal with the "kangaroo problem," which complicates the object-ground-movement computations of an AV, or navigating urban environments as in the Grand Challenges, the obstacles and corner cases that perplex researchers and developers most are not only those that confuse the AV's sensors but also those that involve unpredictable relationships. AV researchers have come to realize that the only way forward in achieving anything like level five full automation is to construct AVs' autopilot systems as part of an ecosystem and whole infrastructure of communication (vehicle to vehicle, vehicle to infrastructure, vehicle to environment). This is not always the case with commercial developers – like Tesla – embedded as they are in competition for market share. Ownership of the means of data production and processing is crucial in this struggle. Tesla, for instance, has placed such importance on effective machine vision because it allows a level of autonomy of operation, less reliance on infrastructural machine communication. This creates a scenario that to a large extent shuts down the social learning needed not just for improving safety through openness and transparency but also for effectiveness if AVs are to ever learn a comprehensive level of machine vision awareness.

Both *how* an AV sees and how it *learns* have become the battleground for pushing the limits and the application of machine vision out into the world. Part of this politics involves an intimate awareness of that machine vision, a camera consciousness, from drivers who have to know its limits and be prepared to take over at any instance and from other road users who need to know an AV's capacities for awareness, their intentions and actions in order to make adequate adjustments. The automated awareness and response capacities of an AV must be matched by an infrastructural and human consciousness of those capacities, an achievement that seems a way off, in part because of the tendency toward proprietary development practices and an inherent dualism that positions humans (as deficient) in strict opposition to vehicles. In the final chapter we consider the sense in which both human and machine "literacy" will play a role in shaping the beneficial integration of machine vision in our everyday lives, and social and urban contexts.

Notes

1. Jake Evans, "Driverless Cars: Kangaroos Throwing Off Animal Detection Software," *ABC News*, June 24, 2017, www.abc.net.au/news/2017-06-24/driverless-cars-in-australia-face-challenge-of-roo-problem/8574816; Allana Aktar and Jalopnik, "Volvo's Driverless Cars Can't Figure Out Kangaroos," *Gizmodo*, June 27, 2017, www.gizmodo.com.au/2017/06/volvos-driverless-cars-cant-figure-out-kangaroos/.
2. AAMI, "Animal Car Accidents Data 2019," *AAMI*, May 27, 2019, www.aami.com.au/aami-answers/insurancey/aami-reveals-peak-periods-for-animal-collisions.html.
3. Soumya Kanti Datta, et al., "Integrating Connected Vehicles in Internet of Things Ecosystems: Challenges and Solutions," in *Proceedings of the 2016 IEEE 17th International Symposium on A World of Wireless, Mobile and Multimedia Networks (WoWMoM)* (New York: IEEE, 2016), 1–6); Ning Lu, et al., "Connected Vehicles: Solutions and Challenges," *IEEE Internet of Things Journal* 1, no. 4 (2014): 289–299; Elisabeth Uhlemann, "Introducing Connected Vehicles [Connected Vehicles]," *IEEE Vehicular Technology Magazine* 10, no. 1 (2015): 23–31; Jack Stilgoe, "Machine Learning, Social Learning and the Governance of Self-driving Cars," *Social Studies of Science* 48, no. 1 (2018): 25–56.
4. Juan Carlos Torres, "Two New Systems Can Help Driverless Cars to See Better," *Slash Gear*, December 23, 2015, www.slashgear.com/two-new-system-can-help-driverless-cars-see-better-23419658/.
5. U.S. Department of Transport (2018). Preparing for the Future of Transportation: Automated Vehicles 3.0. US, www. transportation. gov/av/3.
6. Ibid., vi.
7. Ibid.
8. John R. Quain, "What Self-driving Cars See," *New York Times*, May 25, 2017, www.nytimes.com/2017/05/25/automobiles/wheels/lidar-self-driving-cars.html.
9. See, for example: Ryan Whitwam, "How Google's Self-Driving Cars Detect and Avoid Obstacles," *ExtremeTech*, September 8, 2014, www.extremetech.com/extreme/189486-how-googles-self-driving-cars-detect-and-avoid-obstacles; Jack Stewart, "Why Self-Driving Cars Need Superhuman Senses," *Wired*, September 11, 2017, www.wired.com/story/why-self-driving-cars-need-superhuman-senses/.
10. Stilgoe, "Machine Learning."
11. Luminar Technology, www.luminartech.com/technology/index.html.
12. J. P. Mangalindan, "Uber Acquires Mapping Startup deCarta," *Mashable Australia*, March 4, 2015, http://mashable.com/2015/03/03/uber-acquires-mapping-decarta/#_hunXWPpFaqF; Aaron Souppouris, "The German Car Industry Is Buying Nokia's Here Maps," *Engadget*, August 3, 2015, www.engadget.com/2015/08/03/nokia-here-maps-sale-bmw-mercedes-audi/.
13. Mark Prioleau, "Intel Pays $15B for Mobileye: A Strategic Play for Data," *Medium*, March 14, 2017, https://medium.com/@mprioleau/intel-pays-15b-for-mobileye-a-strategic-play-for-data-672eacd8bb7a.
14. Ibid.
15. Rowan Wilken and Julian Thomas, "Maps and the Autonomous Vehicle as a Communication Platform," *International Journal of Communication* 13 (2019), https://ijoc.org/index.php/ijoc/article/view/8450.
16. See: "PyTorch at Tesla – Andrej Karpathy, Tesla," YouTube video, 11:10, "PyTorch," November 6, 2019, www.youtube.com/watch?v=oBklltKXtDE&t=47s.

17. Stephanie Mlot, "New Systems Teach Driverless Cars to 'See'," *PC Mag*, December 24, 2015, http://au.pcmag.com/software/40861/news/new-systems-teach-driverless-cars-to-see.

18. Ibid.

19. Vijay Badrinarayanan, Alex Kendall, and Roberto Cipolla, "Segnet: A Deep Convolutional Encoder-Decoder Architecture for Image Segmentation," *IEEE Transactions on Pattern Analysis and Machine Intelligence* 39, no. 12 (2017): 2481–2495.

20. Mariusz Bojarski, et al., "End to End Learning for Self-Driving Cars." *arXiv preprint arXiv:1604.07316* (2016), https://arxiv.org/abs/1604.07316.

21. James J. Gibson, *The Ecological Approach to Visual Perception* (London: Lawrence Erlbaum, 1986).

22. Evans, "Driverless Cars."

23. Ian Hutchby, "Technologies, Texts and Affordances," *Sociology* 3, no. 2 (2001): 444, 441–456. See also: Sandra K. Evans, et al., "Explicating Affordances: A Conceptual Framework for Understanding Affordances in Communication Research," *Journal of Computer-mediated Communication* 22, no. 1 (2017): 35–52.

24. Gibson, *The Ecological Approach to Visual Perception*, 34.

25. Maurice Merleau-Ponty, *Phenomenology of Perception*, trans. C. Smith (New York: Routledge, 1962 [1945]).

26. Gibson, *The Ecological Approach to Visual Perception*, 121.

27. Chris Urmson, quoted in Kara Swisher and Chris Urmson, "Full Transcript: Self-driving Car Engineer Chris Urmson on Recode-Decode," *Recode*, September 8, 2017, www.recode.net/2017/9/8/16278566/transcript-self-driving-car-engineer-chris-urmson-recode-decode.

28. Deborah Braid, Alberto Broggi, and Gary Schmiedel, "The TerraMax Autonomous Vehicle," *Journal of Field Robotics* 23, no. 9 (2006): 706.

29. Urmson, quoted in Swisher and Urmson, "Full Transcript."

30. Andre Bacha, et al., "Odin: Team VictorTango's Entry in the DARPA Urban Challenge," *Journal of Field Robotics* 25, no. 8 (2008): 473.

31. Michael Montemerlo, et al., "Junior: The Stanford Entry in the Urban Challenge," *Journal of Field Robotics* 25, no. 9 (2008): 596.

32. Anand R. Atreya, et al., "Prospect Eleven: Princeton University's Entry in the 2005 DARPA Grand Challenge," *Journal of Field Robotics* 23, no. 9 (2006): 746.

33. Chris Urmson, et al., "Autonomous Driving in Traffic: Boss and the Urban Challenge," *AI Magazine*, Summer (2009): 20; Bacha, et al., "Odin," 469.

34. Jonathan Bohren, et al., "Little Ben: The Ben Franklin Racing Team's Entry in the 2007 DARPA Grand Challenge," *Journal of Field Robotics* 25, no. 9 (2008): 607–608.

35. Anna Petrovskaya and Sebastian Thrun, "Model Based Vehicle Detection and Tracking for Autonomous Urban Driving," *Autonomous Robots* 26 (2009): 127.

36. Aarian Marshall, "Why Self-driving Cars *Can't Even* With Construction Zones," *Wired*, February 10, 2017, www.wired.com/2017/02/self-driving-cars-cant-even-construction-zones/?mbid=social_gplus.

37. Ibid.

38. Turker Özkan, et al., "Cross cultural Differences in Driving Behaviours: A Comparison of Six Countries," *Transportation Research Part F: Traffic Psychology and Behaviour* 9, no. 3 (2006): 227–242.

39. Bohren, et al., "Little Ben," 613.

40. Jussi Parikka, *Insect Media: An Archaeology of Animals and Technology* (Minneapolis: University of Minnesota Press, 2010).

41. Ernst D. Dickmanns, Reinhold Behringer, Dirk Dickmanns, Tobias Hildebrandt, Markus Maurer, Frank Thomanek, and Joachim Schiehlen, "The Seeing Passenger Car 'VaMoRs-P'," *Proceedings of the IEEE Intelligent Vehicles Symposium*, 1994: 68–73. doi:10.1109/IVS.1994.639472.

42. "Autonomous Vehicle Recognition Through Localization in 6 Degrees of Freedom (6DoF)," YouTube video, "Civil Maps," May 25, 2017, https://youtu.be/O6DRfAC1JXA.
43. Bohren, et al., "Little Ben," 612.
44. Civil Maps, "Autonomous Vehicle Recognition."
45. Anna Petrovskaya and Sebastian Thrun, "Model Based Vehicle Detection and Tracking for Autonomous Urban Driving," *Autonomous Robotics* 26 (2009): 123–139.
46. Civil Maps, "Autonomous Vehicle Recognition."
47. Parikka, *Insect Media*, 170.
48. Ibid.
49. Stilgoe, "Machine Learning," 31.
50. Ibid., 43.
51. Tesla, "A Tragic Loss," *Tesla*, June 30, 2016, www.tesla.com/blog/tragic-loss.
52. Ibid.
53. Stilgoe, "Machine Learning," 34.
54. Ibid., 28; Albert Bandura, "Organisational Applications of Social Cognitive Theory," *Australian Journal of Management* 13, no. 2 (1988): 275–302; John Dewey, *Essays in Experimental Logic* (Chicago, IL: University of Chicago Press, 1916). Frank Fischer, *Citizens, Experts and the Environment: The Politics of Local Knowledge* (Durham, NC: Duke University Press, 2000). Brian Wynne, "Risk and Social Learning: Reification to Engagement," in *Social Theories of Risk*, S. Krimsky and D. Golding (Eds.) (Westport, CT: Praeger, 1992), 275–300.
55. University of Michigan Transportation Research Institute (UMTRI) http://umtri.umich.edu/our-focus.
56. Rob Matheson, "Helping Autonomous Vehicles See Around Corners," *MIT News*, October 27, 2019, http://news.mit.edu/2019/helping-autonomous-vehicles-see-around-corners-1028.
57. Rob Matheson, "Better Autonomous 'Reasoning' at Tricky Intersections," *MIT News*, November 4, 2019, http://news.mit.edu/2019/risk-model-autonomous-vehicles-1104.
58. Eva Fraedrich and Barbara Lenz, "Societal and Individual Acceptance of Autonomous Driving," in *Autonomous Driving*, Markus Maurer, J. Christian Gerdes, Barbara Lenz, and Hermann Winner (Eds.) (Berlin: Springer, 2016).
59. "Machine Learning," 45. Harry Surden, and Mary-Anne Williams, "Technological Opacity, Predictability, and Self-Driving Cars," *Cardozo Law Review* 38 (2016): 121–181. See also Lance Eliot, "Human Drivers Bullying Self-Driving Cars: Unlawful or Fair Game?" *Forbes*, June 4, 2019, www.forbes.com/sites/lance eliot/2019/06/04/human-drivers-bullying-self-driving-cars-unlawful-or-fair-game/#a8f697e49eed.
60. Stilgoe, "Machine Learning," 41.
61. See, for example: "Tesla Model X | Full Self Driving incl. Tesla Vision Feed," YouTube video, 3:28. "Teslafinity – Sustainable Progress," November 19, 2019, www.youtube.com/watch?v=9ydhDQaLAqM.
62. Barry Brown and Eric Laurier, "The Trouble with Autopilots: Assisted and Autonomous Driving on the Social Road," *CHI 2017*, Denver, CO, May 6–11, 2017, 416–429.

7

AUTOMATING VISUAL LITERACIES

How do machines make sense of the things that they see? This is a question that has preoccupied artist David Rokeby and forms the focus of his long-running work *Giver of Names* (1991–). In one iteration of this work, staged in 2000, the gallery installation consisted of an empty pedestal, a video camera, a computer system and a video projection facility. Visitors to the exhibition would choose from an array of everyday objects (a cone, toy cars), which were strewn across the gallery floor as if it were a child's play space, and then arrange their selection on the pedestal. A video camera would then capture vision of the items placed on the pedestal, relaying this vision to the computer system. The computer would first seek to interpret the arranged objects, looking at each item in turn (distinguishing them by shape, outline, color) and then at the relations between them (see Figure 7.1).[1]

Once defined and identified, the computer would then draw from a stored textual corpus to create, through "free association," a sentence about the pedestal-mounted assortment of items. The resulting sentence would then be spoken by the computer system, and displayed via video projection on a screen, with font colors for the sentence matching those of the arranged objects. For Rokeby, *Giver of Names* "is an exploration of the various levels of perception that allow us to arrive at interpretations, and creates an anatomy of meaning as defined by associative processes."[2] Computer perception, even in this early model, can be understood as relational and environmental; it happens in an embedded context, through decisions made on the basis of the interaction of things, sensors and meaning.

As we have illustrated throughout this book, interest in seeing machines has grown rapidly since the early iterations of Rokeby's project. The computer science has developed exponentially, as has the hardware – digital cameras,

FIGURE 7.1 David Rokeby, *The Giver of Names* at the Art Gallery of Windsor, 2008.
Source: Photo: David Rokeby.

chipsets and processing units. What remains perplexing, and to some extent deeply problematizes attempts to build automated vision systems and find applications for them in the world, is the underlying configuration of sensemaking or meaning-making that these devices are continually enacting. A much longer tradition of thought and research has gone into understanding the complexities of images and the processes of meaning-making in relation to them. As Kate Crawford and Trevor Paglen argue in their critique of face recognition systems and image datasets:

> Images are remarkably slippery things, laden with multiple potential meanings, irresolvable questions, and contradictions. Entire subfields of philosophy, art history, and media theory are dedicated to teasing out all the nuances of the unstable relationship between images and meanings.[3]

And yet there is little work connecting these traditions with the instrumental frameworks for understanding visual meaning-making that is built into computer vision systems. Returning to a key theme raised at the outset of this book, and bubbling away amongst its case studies, we are only at the beginning of the effort required to build the new kinds of literacy needed to both comprehend and intervene in the automated visual systems that are embedding themselves in our everyday lives.

In this final chapter, we bring a literacy-based approach to addressing the big questions of machine learning regarding automated awareness in order to draw together the threads of the different contexts, devices and applications covered in this book. One core question our case studies have raised is: how do people become adequately digitally, visually and data literate in the face of camera conscious seeing machines? Another equally interesting and important question this book implicitly raises is: what kind of literacy is required for an automated vision system, a seeing machine, to be truly aware of its environment and able to adequately act within it? This second question has a long history. As noted in Chapter 5 in relation to drone vision, John McCarthy posited in a 1969 paper with Patrick Hayes that "a computer program capable of acting intelligently in the world must have a general representation of the world in terms of which its inputs are interpreted."[4] The visual literacy at the heart of this challenge is actually shared between humans and intelligent machines. Understanding the facets of the process of generating a "general representation of the world" and an ability to *interpret* inputs remains the core target of ongoing work in AI and machine vision.

Our wager is that the next phases of machine vision and AI development and application will by necessity be built on new visual literacies shared by the humans designing these systems and affected by them, and by the machines. We illustrate here that machine learning, and associated functions such as transfer learning, and the automated vision systems they power, are already operating within a semiotic framework with both the successes and the failures that come with sensemaking, even for humans. The next step is to better train the trainers and designers of these systems, the image sets and machine learning techniques that combine to computationally make sense of world in machinic acts that resemble McCarthy's account of a programmed machine acting intelligently in the world. If, as we have attempted to illustrate throughout this book, the sensemaking, semiotic and learning processes can be adequately opened to scrutiny and collective input – if they can be socialized – then there is a chance for embedding an ethics of care, responsibility, accountability and social good within the schema of machinic intelligence. The first step, and final part of our book, is to account for what a nonhuman-centric digital literacy might entail.

We are pointing to literacy in machine learning not to turn away from the important role of regulation and structural controls over corporate and state uses of machine vision and in the gathering of intimate biometric human data it relies on and produces. There are further regulatory, ethical and political economy critiques to be made. Our aim is to unlock analysis and research around the ways machines and people are deeply involved in each other's learning to see. Visual literacy is multidirectional. It can account for how machines are made to see, and how they remake seeing and re-envision the social. Our shared, evolving camera consciousness is an underacknowledged but, as we have argued throughout the book, significant site of intervention and redress to those commercial or even authoritarian influences.

Rewiring Visual Literacies

Looking back to the visual anthropology of Gregory Bateson and Margaret Mead, it's easy to see the attraction in using photographic images to build a profile of the character of a people considered in the 1940s as exotic.[5] An image, especially when set up so as to generate natural and objective visual data, presents a seemingly effortless primary interpretation. The anthropologist can then set out all the secondary interpretations that add up to an account of the character of the people and culture being studied. However, in the tradition of visual anthropology, as distinct from documentary filmmaking, the processes of data selection, "scientific recording" and exposition occur by design. As noted in Chapter 2, not only did Bateson and Mead use camera techniques such as an angled viewfinder for "when the subject might be expected to dislike being photographed at that particular moment,"[6] a series of calculations were in place to ensure that explanation and meaning were paired with the footage of the mundane and regular aspects of the Balinese people's everyday lives. They explain the importance of note-making during filming and photographing, because "the photographic sequence is almost valueless without a verbal account of what occurred."[7] And, they note some of the "major assumptions" that guided data capture and selection, such as "that parent-child relationships and relationships between siblings are likely to be more rewarding than agricultural techniques."[8]

As any human operator of a CCTV system knows well, *selection* in a vast ocean of visual data is a vital and challenging step in making sense of a scene in which something of interest may or may not have happened. While Bateson and Mead made selection decisions in relation to their existing theories and schemas, they were also constrained by the scarcity of filmstock. The photographer had to identify a setting that was interesting, but after that "the best results were obtained when the photography was most rapid and almost random."[9] We recount these moments of early visual anthropology to establish the link between the problems of capturing, selecting and interpreting visual knowledge they share with the problem-field of machine vision today.

The work that anthropologists do in moving between visual evidence and theory or conceptualization to develop generalizations, interpretations and explications of a scene are analogous to the work a machine vision system has to achieve. Deep learning and neural nets work to interpret the world – or, more specifically, the datasets that serve as their inputs – to cluster and classify and otherwise identify the connections between unlabeled data according to similarities. These systems map inputs to outputs through statistically determined approximations. Technically, they learn to approximate some sort of unknown function in a dataset: $f(x) = y$. This is about finding the right function or working out the right way to transform x into y to understand their relationship.

Does a computer vision system need a working theory of signs to interpret the world it encounters? Perhaps not directly, but it will invariably be working on the basis of one. It's useful to imagine a machine vision system acting like a visual anthropologist, or a qualitative researcher, trying to make meaning from the inputs received from the world around it. These could be very specific inputs with a precise target – faces within a crowd, for instance. Or it could be a vast array of possible objects on a road or in a flight path that need to be interpreted and behaviors calculated in order to make machine adjustments to avoid collision. In each case, semiotics is at play.

There is a faint buzz of discussion about the connection between semiotics and machine learning among computer scientists seeking to improve the way their systems make meaning from the world they are set up to engage with.[10] Semiotics has been an important method for linguistics, media, cultural and communication studies. In the broadcast era, mass media communication, advertising and marketing developed into a social science of manipulating signs and predicting and hence shaping interpretations. As a means for understanding and intervening in the work that AI systems do in interpreting and acting in the world, semiotics has more to offer. It provides some clues about the changing dynamics of a shared camera consciousness between intelligent machines and their human subjects, designers, operators and programmers.

While there are other semiotic approaches that might be drawn on here, Peircean semiotics is useful because it applies to any sign, including nonlinguistic signs, and involves a dynamic and relational tripartite process of meaning-making in action. In C.S. Peirce's semiotics, meaning-making happens actively in the relationship between (a) a sign (sometimes referred to as a sign vehicle), which could be a word, an image, or even a pixel or group of pixels within an image; (b) an object, which could be a thing in the world or an abstract idea or concept; and (c) an interpretant, perhaps best understood as a combination of an interpretation and the interpreter. Importantly, for the machine learning analogy, the interpretant does not have to derive from a person and can combine in chains of meaning-making that lead to new and generalizable interpretants.

The semiotic process, in which an interpretant becomes the sign for another iteration of meaning-making, also describes the layering of processing at work in deep learning. In deep learning networks, data pass through a number of node layers in a multistep process involving a feature hierarchy. When interpreting a digital image, edges are defined first, followed by shapes, followed by objects, arriving at linguistic representations or interpretations of complex objects like faces. Continuing the layering of machine learning knowledge production, transfer learning solves another problem, which is related to how attention, selection and memory work to build interpretations from new data on the basis of previous calculations. In the same way that semiotic meaning-making builds meaning from chains of sign-object-interpretants, transfer learning takes a model developed for one task as the starting point for developing a model applied to a new task.[11]

Selection, attention, detection of similarity and anomaly and transfer learning are all core features of a machine learning semiotic system. What's interesting and important about these processes for understanding the implications of the proliferation of machine learning and machine vision is the tension embedded within the interpretative act. As Iddo Tavory and Stefan Timmermans argue in their theorization of semiotics in qualitative research, the dynamic movement from an encounter with a sign-object toward an interpretation or theoretical generalization involves "a tension between our ability to imbue the world with meaning and the world's resistance to such attempts."[12] In and around the calculations that go into machine learning systems are sets of parameters, weightings, assumptions and expectations that help to guide and shape interpretations. Whether establishing the labeled training datasets to guide the outputs of a supervised machine learning system or setting up the optimal algorithms and parameters for an unsupervised system to detect patterns in unlabeled data, a new kind of visual literacy is beginning to take shape. Visual literacy, in other words, can be understood against the grain to describe the "competency" of an automated machine vision system as it both interprets and produces elements of the visible world as inputs and outputs through recognizable semiotic processes.

Traditionally, literacy has been defined around a child's ability to read and write linguistic texts. It also refers to a competency in a particular area, and as systems of communication expanded into media and digital media domains, new fields of media literacy, digital literacy, visual literacy and data literacy have taken shape. As an umbrella field of enquiry, digital literacies is often pluralized to reflect the individual and variable nature of the way we encounter and engage with all of the possible objects of the digital world, and the different sets of competencies, this might involve. It refers to an emerging field oriented toward education and empowerment in which people are faced with adapting to ever changing digital environments.[13] Rodney Jones and Christoph Hafner see digital literacies as an adaptive set of abilities, skills and knowledge about the operation, use and cultures of digital media and networked systems.[14] The strength (and for some frustration) of Paul Gilster's 1997 account in his concept-defining book *Digital Literacy* was its focus on the shifts brought about by web environments as they shaped information, interactions, transactions and consumption.[15] Addressing the implications of web browsers, hyperlinks and the functions of websites, Gilster's work highlighted the need for all to take action to ensure their ability to capitalize on what these new technologies had to offer. The point of taking a literacies approach for Gilster and others, and for the new environment of AI and machine vision systems, is to both aid understanding of these socio-technical shifts and help people take advantage of the opportunities and possibilities they create.

Visual literacy is the field of education practice and research established to help young people navigate the rich multimodal and highly visual social environment of contemporary digital life. In the 1960s, around the time many of the techniques of image processing and early machine vision technologies were being thought up, John Debes offered the first definition of visual literacy focusing on the human competencies that enable a person to "discriminate and interpret the visible actions, objects, symbols" they encountered in their environment.[16] Eva Brumberger writes about the "myth of visual literacy" as something that people, and specifically young people, in the digital age "naturally" possess. The visual thinking, learning, interpretive and productive elements of visual literacy are also encompassed by cultural, political and historical contexts.[17] Brumberger's work is a reminder of the increasing complexity and necessity of visual literacy in the age of digital media – the complexity and necessity of which is even greater in the age of automated media.

Most contemporary approaches to digital literacy take a sociocultural view of literacy "as a set of socially organised practices."[18] This means not just knowing how to encode and decode in a particular kind of script but also adapting to the contexts and purposes of the environment of language use or interactions. Attention to the consequences of machine vision systems, as we have addressed throughout this book, regarding issues such as bias, unfairness, misrepresentation or risk in inaccuracy, has happened reflexively, often through agitation and investigation through technology reporting, through algorithmic auditing and critical algorithm studies, through activist art projects and, in the case of the face recognition technologies discussed in Chapter 3, through political pushback and emergent regulation. These efforts can be better armed through a literacies approach and an understanding of the semiotic processes in which machine vision systems engage.

All of this is to say that we need to better socialize machine vision and machine learning to aid its competency and guide its operation in the world toward social good – as a form of accountable, responsible machinic literacy. Before returning to the case studies covered throughout this book, it is worth exploring one of the startling applications of machine learning and image processing that has come to illustrate the new machinic literacy and complex negotiations of the semiotic in the age of machine vision: deepfakes.

It Cuts Both Ways: Making and Detecting Deepfakes

The tensions and possible trajectories of state-of-the-art contemporary machine vision systems have been cast into the public eye by spectacular developments in the production of *deepfakes*. Deepfakes are a synthesis of digital video and audio achieved through deep learning techniques, producing an artificially constructed video that looks like the real thing. The same deep learning techniques used in the face swapping apps discussed in Chapter 3 are used to replace a face in one video with another, provided there is a sufficient dataset to draw on. There is much to be said about the social implications of undetectable videos of politicians and other public figures, or image-based abuse through deepfake pornographic videos. Deepfake videos have emerged at a time when concerns have already resurfaced about the urgent need for media literacy in the wake of scandals involving "fake news," and the political interventions and abuse of social media bots and other forms of automated and invasive intervention into digital media environments for political or commercial gain. Deepfake videos have been described in this context as a mechanism for "democratizing fraud," with significant consequences for social trust in visual information.[19] For our purposes, however, deepfakes put to the test some of the core components of camera conscious seeing machines – their capacity for a new machinic visual literacy, in both the read and write senses of that term.

Because of their ability to intervene in the visible, to generate visual data artificially, deep learning systems are beginning to disrupt the objectivity conceived to be present in video and audio evidence. But we've been here before. Digital photography and its manipulations disrupted the boundaries of visual evidence in photography more than a decade ago.[20] As many scholars have argued, and a report on deepfakes by the Data & Society Institute reiterates, "the 'truth' of audiovisual content has never been stable – truth is socially, politically, and culturally determined."[21] What is being tested now is more-than-human vision and perception; it is also the capability of machine vision systems to discriminate authentic from artificial, their ability to *generate* visual material (or truths) and to outsmart themselves in encoding and decoding moving images. In fact, the problems of accountability with deepfakes stem from their underlying technology – Generative Adversarial Networks (GAN).

Introduced in 2014,[22] GANs are deep neural net architectures in which two deep learning models are pitted against one another. A *generator* network creates new data instances, usually based on set parameters or training data, while another, the *discriminator*, evaluates them, attempting to identify anomalies against the training dataset, in turn pushing the generator to create more accurate outputs to "fool" the discriminator. As AI infrastructure start-up Skymind puts it: "The goal of the generator is to generate passable hand-written digits: to lie without being caught. The goal of the discriminator is to identify images coming from the generator as fake."[23] This means that, as they generate realistic fakes, GAN systems are designed to exceed their own and other deep neural net systems' capability to detect their inauthenticity. This level of read-write literacy will have many applications beyond political misinformation and artificial pornography, although these areas of social impact are themselves significant enough to provoke political and regulatory action around the world. Many social media platforms moved quickly to find ways to detect and block deepfakes.

Political satire and media arts have taken on the task of building attention around the consequences of deepfakes. One such project, by activist-artist Bill Posters and artist and critical technologist Daniel Howe called Spectre, has used multiple channels to gain a broader understanding, raise critical awareness of and point to the potential social impact of the capacity to manipulate our visual fields through deepfakes. The work premiered on June 6, 2019, at Site Gallery, Sheffield, as part of the Sheffield Doc/Fest "Alternate Realities" exhibition. For the Spectre project, Posters and Howe collaborated with AI technology start-ups, including CannyAI, Respeecher and Reflect, "to create a range of AI-generated 'deep fake' celebrities – including dead and living artists – that push the fields of generative art and digital storytelling to new heights."[24] The opening made waves across social media and featured in submissions to a US Senate hearing on the dangers of deepfake, AI and machine learning technologies.[25] While videos targeted a number of "influencers," it was the deepfake video of Facebook founder Mark Zuckerberg that pushed political buttons and gathered widespread attention. The challenge set by the video, also hosted on Bill Posters' Instagram account, was whether Facebook would take a stand on the use of the technology and remove or hide the videos. To date, they have not.

Interventions like Spectre seek to profile the broader problems involved in the ability of machines to remake the visual and the visible. They point to our increasing reliance on those same machines to extend, correct or detect visual inauthenticity. The questions remaining are who or what needs to be trained? And how?

Training Humans and Machines for the New Camera Consciousness

If machines can be considered increasingly visually literate in the ways described earlier, to the point where they are navigating and mapping their environment (drones, cars, mobile phones), or discerning objects, people and actions in public spaces (CCTV and face recognition), or even generating new components in the visual field (deepfakes), there is a pressing need to isolate the key sites for intervention. Critical attention has been directed at the datasets from which machine learning algorithms draw. As noted in Chapter 3, machine learning took off with the accessibility of very large datasets problematically brought about by shared online content. All of the applications for machine vision covered in this book, and those yet to be developed, rely on the strength and accurate functioning of training datasets that establish the "ground truths" for successful automation. The end-goal of an inclusive AI, its application for social good impact, is bound up in the ethics of care and social inclusion embedded in the processes of developing and working with AI datasets and system design. These are both human and machine processes. Our argument is not simply that it is up to benevolent humans to ensure that bias and inaccuracy are overcome in machine learning AI. Humans do not have a great track record or infallible aptitude when it comes to judgment on the basis of seeing. In fact, many (although not all) of the biases found in AI systems come from human judgment in the process of image labeling or parameter setting.

While developing new machine learning techniques and innovative algorithms for problem-solving, AI researchers have also focused heavily on the well-known problem of imbalance and bias in datasets.[26] These are not easy problems to solve. Datasets have always been resource-intensive to produce and to accurately label so as to be useful benchmarks for identifying patterned similarities or anomalies in new data. Imbalance and bias can occur at the point of data collection, for instance in the use of face images or video of celebrities, where racial, gender, age or other physical features may be unbalanced or missing, and so not as easily recognized or detected by systems trained on that set. It can also occur in the process of labeling images. In a semiotic schema, labels are both a sign vehicle, standing in for the object in question (the original sign vehicle most likely being a digital image), and an interpretant. That is, most labeling for the datasets used for supervised machine learning is generated manually, often through crowdsourced processes such as Amazon's Mechanical Turk (more on that provided later). This process of manually labeling data enacts many instances of interpretation, making labels the new functional sign that results from human interpretation.

A lot of human labor goes into training machines to become visually literate. And a lot can go wrong, and the stakes are high. As Crawford and Paglen explain, "These datasets shape the epistemic boundaries governing how AI systems operate, and thus are an essential part of understanding socially significant questions about AI."[27] They are often built on "shaky and skewed assumptions."[28] Crawford and Paglen go on to remind us that

> The circuit between image, label, and referent is flexible and can be reconstructed in any number of ways to do different kinds of work. What's more, those circuits can change over time as the cultural context of an image shifts, and can mean different things depending on who looks, and where they are located. Images are open to interpretation and reinterpretation.[29]

Not unlike the processes qualitative researchers use to "code" datasets drawn from interviews, observations, videos and other research interactions, AI datasets, like the 14 million item Stanford University built ImageNet dataset, are labeled through a combination of taxonomy, category and subcategory and class. Labeling is by design reductive and standardized, like any metadata attribute.[30] This act of simplification is where the slippages, inconsistencies, inaccuracies, biases and politics come into play.

Crowdsourced image labeling raises questions of expertise and bias among the massive workforce that has amassed to undertake the work at low labor cost. Amazon's Mechanical Turk (MTurk) has been the principal site for organizing crowdsourced labeling for machine learning datasets. Describing itself as a "crowdsourcing marketplace," MTurk courts machine learning and AI developers building labeled machine vision datasets. The widespread use of the service has also led to research examining the benefits and problems in using such a distributed workforce,[31] as well as scholarship critical of associated labor practices.[32] A related line of research in developing labeled training datasets for machine learning purposes has explored possibilities for "weak-supervision."[33] Weak supervision promises to offer a way of injecting domain expertise back into the "black box" processes of representational deep learning systems. It is an intervention into the semiotic chain, so that, rather than producing labels for image data for instance, domain expert labelers generate "labeling functions" that establish the parameters, rules or heuristics for generating label sets.

The struggle to extend, rewire and automate visual literacy is well underway and is driving developments in the field. In her book on AI, Ellen Broad points to the critical work of a concerned scientist blogging about the inaccuracies built into a public image dataset of chest X-ray images (ChestX-ray8) released by the US National Institutes of Health (NIH) Clinical Centre (2018).[34] The aim of making the dataset publicly accessible was to encourage researchers to develop systems to detect and diagnose disease. When Australian radiologist and machine learning PhD researcher Luke Oakden-Rayner examined random samples of the X-rays labeled with different classes of disease, he found that many of them were mislabeled, and he declared in his blog that the dataset was not fit for use.[35] Domain expertise of this sort is an issue for dataset labeling where there is any degree of specialist knowledge needed. What Oakden-Rayner found in this case is the uncertainty inherent in the semiotic chain that generates the baseline or ground truth for machine vision systems.

The ChestX-ray8 dataset, consisting of 108,984 X-rays from 32,717 patients collected between 1992 and 2015, was developed using weak-supervision classification methods (Figure 7.2).[36] The process, as explained by Wang et al., involved text-mining eight common disease labels from the associated text radiological reports, and then using weak-supervision techniques to demonstrate that "these thoracic diseases can be detected and spatially located via a unified weakly-supervised multi-label image classification and disease localization formulation."[37] This process then served to apply classifications in a way that bypasses the need for manual labeling of the large dataset by labelers with clinical expertise. Whether Oakden-Rayner was right in his critical assessment of the fitness of the dataset for use in medical machine vision, what we are

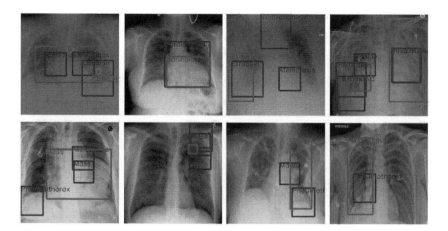

FIGURE 7.2 Labeled medical images; extract from the ChestX-ray8 dataset.

seeing here is the battle at the edges of human–machine interaction in the AI-based extension of visual literacy.

Enhancing Camera Consciousness: Conclusions

Building machine vision systems involves activating and training camera consciousness as a new hybrid human–machine sensibility and literacy. The complexity of machine learning and AI generally can render it opaque and resistant to broader input outside of the computer and data science developers and the big tech companies. However, the literacy-based approach to AI explored in this chapter encourages practices of recognizing, interrogating and seeking to affect each of the steps in the semiotic chains through which those systems operate. It opens the door to social scientists, activists and citizens alike to participate in system design for social good. The sign–object–interpretant relationships and chains of meaning-making can easily go awry. They can cause exclusions, increase vulnerabilities, inequalities, accidents and intrusions; but they can also be trained for social good, to identify and help correct social imbalance and inequality, improve objectivity or fairness in automated decision-making.

Critical accounts of AI systems and their imposition of new forms of control and power are essential to contemporary digital citizenship – a process that takes form in rights claims and individual and collective acts of resistance to applications of those systems that impact on everyday freedoms.[38] While in this book we have only sought to illustrate the shared human–machine involvement in camera awareness, we hope this perspective and analysis can help turn these systems toward redressing inequality, improving health and well-being, improving cities and generating new visibility for social inclusion purposes. This, we would suggest, must become an explicit goal in the form of an inclusive AI.

And so, we return to the book's content and what we set out to achieve with each chapter and with the book as a whole. In the opening chapter of the book, we established the overall terrain of the book and gave an abbreviated account of the core concept (camera consciousness) and its key thematic concerns (the social impacts of machine vision, the importance of visual literacies). As we explained in that chapter, camera consciousness involves both an awareness of cameras *and* a self-conscious reaction to their power to make visible, to reveal, capture and hold their target, and to fix it for scrutiny, judgment and action.

Chapter 2 further developed the concept of camera consciousness, positing it as an effective way of accounting for the social changes involved with the increasing autonomy and intelligence of camera technologies and image processing. We began with the visual anthropology of Gregory Bateson and Margaret Mead and used the concept of camera consciousness to develop an analytical framework, with the aid of William James, Gilles Deleuze, Jane Bennett and other contemporary theorists. The chapter considered the "ways of seeing" and altered visuality of new "intelligent" camera technologies and explored the growing value of visual data. A central claim of this chapter is that a new camera consciousness is needed now to realize the visual knowledge generated by distributed, mobile, aerial and autonomous cameras and our modes of acting with them. Developing a keener sense and understanding of the new camera consciousness at play is key and involves ongoing social research and technical, cultural, experiential and political-economic elaboration and analysis.

In Chapter 3, we examined the value of face images in the application of face recognition technology. Expanding on the example of China's camera and face recognition-assisted "social credit system" and its "algorithmic governance" goals, this chapter offered an alternative point of focus to a reading that prioritizes surveillance and governance. We focused not on the loss of privacy these systems entail, but, rather, on the new modes of digital inclusion and exclusion, connection and sanction that they introduce. This chapter also explored some of the key technological components of face recognition technology, where the technology has come from, where it is tracking now and where it is headed. Built on technologies of recognition and matching, identification and sentiment analysis, the long-coveted visual data of the face offers a wide range of possible applications. While many of the current applications and future prognostications relating to AI, machine learning and face recognition range from the troubling to the distinctly dystopian, we ask whether it might now be time to begin to explore face recognition for its possible social good applications, including as a form of computational responsiveness to individuals, which could incorporate explicit social inclusion interventions and considerations.

Chapter 4 considered the automating and augmenting of mobile vision capture. The chapter opened with a historical account of key steps and stages in the development of camera automation, and of the incorporation of cameras into mobile phones. Then, in the second part of the chapter, the emphasis shifted. Looking beyond the predominant focus on camera phone practices in studying the application of current mobile media, we turned to a consideration of the autonomous production of information and metadata, including geo-locative metadata, that accompany and are left behind by camera phone and smartphone use, as well as to the relative autonomy of mobile imaging and the increasing value of visual data. The chapter then explored the transition toward a more embedded, everyday activation of augmented reality. Taking Google's Project Tango as a focus, our argument in the second part of the chapter was to view this project as emblematic of the current and likely future trajectory of personal mobile media as involving sophisticated sensing and visual processing power. Through developments like Project Tango, mobiles provoke questions about how individual local environments become visible and socially available, and what the possibilities, and possible implications are, of seeing and acting digitally beyond human senses.

In Chapter 5, we asked: How does the increasingly ubiquitous aerial drone affect and depend on new forms of distributed, wireless visuality? While a lot of attention has been paid to the trajectory of drone technology as it has arisen out of military contexts and has since invaded urban spaces, we examined the essential role played by drones' camera technology, visual controls and wireless relations. To understand the implications of drones (both small and recreational, and large and lethal), we looked at how they have reconfigured personal, public and environmental visibility – or what it means to see and be seen today via drone cameras. Drone vision pushes the boundaries of, and tests new understandings of, what is considered "public."

This chapter also examined the semiautonomous vision technology that allows drones to act in the world, and to control their alternative viewfinder from the skies. It explored the "creepy agency" of drone cameras as a factor of the new visual knowledge they are able to produce, and probes the meaning of "autonomy" for these machines that act *like* insect or animal in their unpredictable movements and their ability to swarm or to take the position of the "fly on the wall," but also their waywardness. From this, we argued that drone vision presents us with a key case of an altered sociality triggered by a highly contested camera consciousness. Given the autonomy of these perceptual systems, again, a new camera consciousness is required, and the question of how we experience and make sense of drone vision continues to be vitally important and remains a live question.

In Chapter 6, we examined a little understood aspect of autonomous vehicle technology: vision capture. While most attention relating to driverless cars has focused on the legal and safety concerns associated with removing a driver's control, fundamental to the long-term success of autonomous vehicles is the development of visual capture and processing – that is, the processes that are involved in the mapping, sensing and real-time dynamic visual data processing within dynamic urban environments. To become properly integrated into urban, suburban or country transport systems, autonomous vehicles have to change or accelerate in response to the way those spaces, and all their component objects, obstacles and subjects, can be seen or rendered visible, mapped and transposed into real-time traversable data.

In examining vision capture and processing by autonomous vehicles, we asked *how* an AV learns to see, and we explored the human and social challenges that visual processing technologies place in the path of autonomous vehicle developers. This chapter also considered the concept of technological affordances, as studied by William Gibson, in relation to a wholly complex ecosystem. Driverless cars stretch the sense in which a machine can be defined by its affordances for seeing and acting in its environment. With this comes a host of ethical and other questions regarding how and what it sees and why and when it acts or, crucially, when it doesn't act.

And, finally, in this the concluding chapter, we have considered the implications of the technologies discussed throughout the book through the lens of visual literacies. This critical lens prompts us to ask what it really does mean to automate each element of the competencies of visual literacy. At the same time, this same lens leads us to question how automation, machine reading and machine learning supersede what we have understood to be the exclusively human domain of visual communication. Taken together, these two lines of questioning lead us to return to and rethink what a nonhuman-centric media studies might look like in relation to the proliferation of autonomous seeing machines and visual data.

Across this book, then, we have examined four different, contemporary seeing machines: face recognition, smartphones, drones and autonomous vehicles. Each of these technologies has been regarded as expressive of the possibilities and the anxieties or uncertainties of what we have termed the *new camera consciousness*. In developing this idea, we have sought to offer a way of making sense of the increasing array of technologies of automated vision, their re-visioning of visibility, via the vehicles that make them mobile and active in the world. We have argued throughout that these new technologies of vision – these seeing machines – evidence a quickening of the datafication of the visual and the rising value of everyday visual data. We have also argued that these new technologies of vision are becoming increasingly automated.

These seeing machines invite us to critically reconsider what counts as (visual) media and how mediation occurs and affects us. Contemporary seeing machines require a critical reorientation – a new camera consciousness – that helps us understand how we conceive of the camera and the terrain of visibility – as lively assemblages invested with intensity, augmentative and transformative power, and as processual and dense with powerful information beyond the surface, within the metadata. With automated vision comes significant disruptions to existing media and communication ecosystems, everyday social practices and visual publics. This book has provided one way into developing a deeper understanding of the disruptions and transformations wrought by these new camera technologies, emerging visual practices and AI-driven social changes.

Notes

1. "The Giver of Names (1991–) by David Rokeby," YouTube video, 5:59. "David Rokeby," December 8, 2006, www.youtube.com/watch?v=sO9RggYz24Q.
2. Cited in Christiane Paul, "Renderings of Digital Art," *Leonardo* 35, no. 5 (2002): 484.
3. Kate Crawford and Trevor Paglen, "Excavating AI: The Politics of Images in Machine Learning Training Sets," *excavating.ai*, September 19, 2019, https://excavating.ai; William J. T. Mitchell, *Picture Theory: Essays on Verbal and Visual Representation* (Chicago: University of Chicago Press, 2007 [1994]).
4. John McCarthy and Patrick J. Hayes, "Some Philosophical Problems from the Standpoint of Artificial Intelligence," in *Readings in Artificial Intelligence*, Bonnie Lynn Webber and Nils J. Nilsson (Eds.) (Burlington, MA: Morgan Kaufman, 1981), 431.
5. Gregory Bateson and Margaret Mead, *Balinese Character: A Photographic Analysis* (New York: New York Academy of Sciences, 1942).
6. Ibid., 49.
7. Ibid., 49–50.
8. Ibid., 50.
9. Ibid.
10. See, for example: Alex Kearney and Oliver Oxton, "Making Meaning: Semiotics Within Predictive Knowledge Architectures," *arXiv.org, arXiv:1904.09023*. For a very early account, see: Dimtry A. Poepolov, "Semiotic Models in Artificial Intelligence Problems," in *IJCAI '75 Proceedings of the 4th International Joint Conference on Artificial Intelligence – Volume 1* (New York: ACM, 1975), 65–70, www.ijcai.org/Proceedings/75/Papers/010.pdf.
11. See, for example: Jason Brownlee, *Deep Learning for Computer Vision: Image Classification, Object Detection, and Face Recognition in Python* (Vermont, Victoria, Australia: Machine Learning Mastery, 2019).
12. Iddo Tavory and Stefan Timmermans, *Abductive Analysis: Theorizing Qualitative Research* (Chicago: University of Chicago Press, 2014), 25.
13. Michele Knobel and Colin Lankshear, eds., *Digital Literacies: Concepts, Policies and Practices* (New York: Peter Lang, 2008); Rodney H. Jones and Christoph A. Hafner, *Understanding Digital Literacies: A Practical Introduction* (New York: Routledge, 2012); Yoram Eshet-Alkalai, "Digital Literacy: A Conceptual Framework for Survival Skills in the Digital Era," *Journal of Educational Multimedia and Hypermedia* 13, no. 1 (2004): 93–106.
14. Rodney H. Jones and Christoph A. Hafner, *Understanding Digital Literacies: A Practical Introduction* (London: Routledge, 2012).
15. Paul Glister, *Digital Literacy* (New York: John Wiley & Sons, Inc., 1997).
16. John L. Debes, "The Loom of Visual Literacy: An Overview," *Audiovisual Instruction* 14, no. 8 (1969): 27.
17. Eva Brumberger, "The Myth of Digital Literacy and Digital Natives," in *Digital Media: Transformations in Human Communication*, 2nd ed., Paul Messaris and Lee Humphreys (Eds.) (New York: Peter Lang, 2017).
18. Colin Lankshear and Michele Knobel, "Introduction: Digital Literacies— Concepts, Policies and Practices," in *Digital Literacies Concepts, Policies and Practices*, M. Knobel and C. Lankshear (Eds.) (New York: Peter Lang, 2008), 1–16, 4.

19. Robert Chesney and Danielle Citron, "Deepfakes and the New Disinformation War: The Coming Age of Post-Truth Geopolitics," *Foreign Affairs*, 98, January/February (2019), www.foreignaffairs.com/articles/world/2018-12-11/deepfakes-and-new-disinformation-war.

20. Fred Ritchin, *After Photography* (New York: WW Norton & Company, 2009); William J. Mitchell, *The Reconfigured Eye: Visual Truth in the Post-Photographic Era* (Cambridge, MA: MIT Press, 1992).

21. Britt Paris and Joan Donovan, *Deepfakes and Cheap Fakes: The Manipulation of Visual Evidence* (New York: Data & Society, 2019), 2, https://datasociety.net/wp-content/uploads/2019/09/DS_Deepfakes_Cheap_FakesFinal-1.pdf.

22. Ian J. Goodfellow, et al., "Generative Adversarial Nets," in *Proceedings in Advances in Neural Information Processing Systems 27 (NIPS 2014)*: 2672–2680.

23. "A Beginner's Guide to Generative Adversarial Networks (GANs)," *Skymind*, https://skymind.ai/wiki/generative-adversarial-network-gan.

24. Bill Posters, "Gallery: 'Spectre' Launches (Press Release)," *Bill Posters*, May 29, 2019, http://billposters.ch/spectre-launch/.

25. Luke O'Neil, "Doctored Video of Sinister Mark Zuckerberg puts Facebook to the Test," *The Guardian*, 12 June, 2019, www.theguardian.com/technology/2019/jun/11/deepfake-zuckerberg-instagram-facebook.

26. Aditya Khosla, et al., "Undoing the Damage of Dataset Bias," in *European Conference on Computer Vision* (Berlin: Springer, 2012), 158–171; Haw-Shiuan Chang, Erik Learned-Miller, and Andrew McCallum, "Active Bias: Training More Accurate Neural Networks by Emphasizing High Variance Samples," in *Advances in Neural Information Processing Systems* (NIPS) 2017, arXiv.org, arXiv:1704.07433; Antonio Torralba and Alexei A. Efros, "Unbiased Look at Dataset Bias," in *Proceedings of the Conference on Computer Vision and Pattern Recognition (CVPR)* (New York: IEEE, 2011), 1521–1528; Foster Provost, "Machine Learning from Imbalanced Data Sets 101," in *Proceedings of the AAAI'2000 Workshop on Imbalanced Data Sets* 68, no. 2000 (New York: AAAI Press, 2000), 1–3.

27. Crawford and Paglen, "Excavating AI."

28. Ibid.

29. Ibid.

30. Jeffrey Pomerantz, *Metadata* (Cambridge, MA: MIT Press, 2015).

31. See, for example: Eyal Peer, Joachim Vosgerau, and Alessandro Acquisti, "Reputation as a Sufficient Condition for Data Quality on Amazon Mechanical Turk," *Behavior Research Methods* 46, no. 4 (2014): 1023–1031; Alexander Sorokin and David Forsyth, "Utility Data Annotation with Amazon Mechanical Turk," in *2008 IEEE Computer Society Conference on Computer Vision and Pattern Recognition Workshops* (New York: IEEE, 2008), 1–8.

32. Kären Fort, Gilles Adda, and K. Bretonnel Cohen, "Amazon Mechanical Turk: Gold Mine or Coal Mine?" *Computational Linguistics* 37, no. 2 (2011): 413–420.

33. Alexander Ratner, et al., "Data Programming: Creating Large Training Sets, Quickly," *30th Conference on Neural Information Processing Systems (NIPS 2016)*, Barcelona, Spain. https://papers.nips.cc/paper/6523-data-programming-creating-large-training-sets-quickly.pdf.

34. Ellen Broad, *Made by Humans: The AI Condition* (Melbourne: Melbourne University Publishing, 2018).

35. Luke Oakden-Rayner, "Exploring the ChestXray14 Dataset: Problems," *Luke Oakden-Rayner* (blog). December 18, 2017, https://lukeoakdenrayner.wordpress.com/2017/12/18/the-chestxray14-dataset-problems/.

36. Xiaosong Wang, et al., "ChestX-ray8: Hospital-scale Chest X-ray Database and Benchmarks on Weakly-Supervised Classification and Localization of Common Thorax Diseases," *Computer Vision Foundation* (2017), 2097–2016. http://openaccess.

thecvf.com/content_cvpr_2017/papers/Wang_ChestX-ray8_Hospital-Scale_ Chest_CVPR_2017_paper.pdf.

37. Ibid., 2098.

38. Engin Isin and Evelyn Ruppert, *Being Digital Citizens* (London: Rowman and Littlefield International, 2015); Anthony McCosker, Son Vivienne, and Amelia Johns, eds., *Negotiating Digital Citizenship: Control, Contest and Culture* (London: Rowman & Littlefield International, 2016).

REFERENCES

AAMI. "AAMI Animal Car Accidents Data 2019." *AAMI*, May 27, 2019, www.aami.com.
au/aami-answers/insurancey/aami-reveals-peak-periods-for-animal-collisions.html.

"A Beginner's Guide to Generative Adversarial Networks (GANs)." *Skymind*, https://
skymind.ai/wiki/generative-adversarial-network-gan.

"About OpenAI." *openai.com*, https://openai.com/about/.

Ackerman, Evan. "Autonomous Vehicles vs. Kangaroos: The Long Furry Tail of Unlikely
Events." *IEEE Spectrum*, July 5, 2017, https://spectrum.ieee.org/cars-that-think/
transportation/self-driving/autonomous-cars-vs-kangaroos-the-long-furry-tail-
of-unlikely-events.

"Aerial Drone Captures Scale of Hong Kong Protests." YouTube video, 0:43. "The Wall
Street Journal," September 29, 2014, www.youtube.com/watch?v=0HoEj1BOOpQ.

Aktar, Allana, and Jalopnik. "Volvo's Driverless Cars Can't Figure Out Kangaroos."
Gizmodo, June 27, 2017, www.gizmodo.com.au/2017/06/volvos-driverless-cars-
cant-figure-out-kangaroos/.

Alquran, Hiam, Isam Abu Qasmieh, Ali Mohammad Alqudah, Sajidah Alhammouri,
Esraa Alawneh, Ammar Abughazaleh, and Firas Hasayen. "The Melanoma Skin
Cancer Detection and Classification Using Support Vector Machine." In *Proceedings of the 2017 IEEE Jordan Conference on Applied Electrical Engineering and Computing Technologies (AEECT), October 11–13*, 1–5. New York: IEEE, 2017, https://doi.
org/10.1109/AEECT.2017.8257738.

"Amazon Rekognition." *Amazon Web Services*, https://aws.amazon.com/rekognition/.

Andrejevic, Mark. *Automated Media*. New York: Routledge, 2019.

Andrejevic, Mark. "Becoming Drones: Smartphone Probes and Distributed Sensing."
In *Locative Media*. Edited by Rowan Wilken and Gerard Goggin, 193–207. London:
Routledge, 2015.

Andrejevic, Mark, and Mark Burdon. "Defining the Sensor Society." *Television and
New Media* 16, no. 1 (2015): 19–36.

Andrejevic, Mark, and Kelly Gates. "Big Data Surveillance: Introduction." *Surveillance &
Society* 12, no. 2 (2014): 185–196.

Armitage, John, and Ryan Bishop, eds. *Virilio and Visual Culture.* Edinburgh: Edinburgh University Press, 2013.

Atreya, Anand R., Bryan C. Cattle, Brendan M. Collins, Benjamin Essenburg, Gordon H. Franken, Andrew M. Saxe, Scott N. Schiffres, and Alain L. Kornhauser. "Prospect Eleven: Princeton University's Entry in the 2005 DARPA Grand Challenge." *Journal of Field Robotics* 23, no. 9 (2006): 745–753.

"Autonomous Vehicle Recognition Through Localization in 6 Degrees of Freedom (6DoF)," YouTube video, 3:19. "Civil Maps," May 25, 2017, https://youtu.be/O6DRfAC1JXA.

Bacha, Andrew, Cheryl Bauman, Ruel Faruque, Michael Fleming, Chris Terwelp, Charles Reinholtz, Dennis Hong, Al Wicks, Thomas Alberi, David Anderson, Stephen Cacciola, Patrick Currier, Aaron Dalton, Jesse Farmer, Jesse Hurdus, Shawn Kimmel, Peter King, Andrew Taylor, David Van Covern, and Mike Webster. "Odin: Team VictorTango's Entry in the DARPA Urban Challenge." *Journal of Field Robotics* 25, no. 8 (2008): 467–492.

Badrinarayanan, Vijay, Alex Kendall, and Roberto Cipolla. "Segnet: A Deep Convolutional Encoder-Decoder Architecture for Image Segmentation." *IEEE Transactions on Pattern Analysis and Machine Intelligence* 39, no. 12 (2017): 2481–2495.

Bakheet, Samy. "An SVM Framework for Malignant Melanoma Detection Based on Optimized Hog Features." *Computation* 5, no. 4 (2017): 1–13.

Bandura, Albert. "Organisational Applications of Social Cognitive Theory." *Australian Journal of Management* 13, no. 2 (1988): 275–302.

"Ban Facial Recognition Map." www.banfacialrecognition.com/map/.

"Bangkok 'Shutdown' 13 Jan 14 Pathumwan." YouTube video, 0:26. "TheCyberJom," January 14, 2014, https://youtu.be/6WWfnRnzZGE?list=PL0PjXgwbkvur0wGByQSVu9r1ODn5TK8m.

Barthes, Roland. *Mythologies.* Translated by Annette Lavers. New York: Hill and Wang, 1972.

Bate, David. "The Digital Condition of Photography: Cameras, Computers and Display." In *The Photographic Image in Digital Culture*, second edition. Edited by Martin Lister, 77–94. New York: Routledge, 2013.

Bateson, Gregory, and Margaret Mead. *Balinese Character: A Photographic Analysis.* New York: New York Academy of Sciences, 1942.

BBC News. "Gatwick Airport: Drones Ground Flights." *BBC News*, December 20, 2018, www.bbc.com/news/uk-england-sussex-46623754.

Bendig, Juliane, Andreas Bolten, and Georg Bareth. "Introducing a Low-cost Mini-UAV for Thermal- and Multispectral-Imaging," *International Archives of the Photogrammetry, Remote Sensing and Spatial Information Sciences* 39, no. B1 (2012): 345–349.

Bennett, Jane. *Vibrant Matter: A Political Ecology of Things.* Durham, NC: Duke University Press, 2009.

Berger, John. *About Looking.* New York: Pantheon, 1980.

Berger, John. *Understanding a Photograph.* New York: Aperture, 2013.

Berger, John. *Ways of Seeing.* London: Penguin, 1972.

Bishop, Ryan, and Jussi Parikka. "The Autonomous Killing Systems of the Future are Already Here, They're Just Not Necessarily Weapons – Yet." *The Conversation*, August 4, 2015, https://theconversation.com/the-autonomous-killing-systems-of-the-future-are-already-here-theyre-just-not-necessarily-weapons-yet-45453.

Blaschke, Estelle. "From Microform to the Drawing Bot: The Photographic Image as Data." *Grey Room* 75 (2019): 60–83.

Bohren, Jonathan, Tully Foote, Jim Keller, Alex Kushleyev, Daniel Lee, Alex Stewart, Paul Vernaza, Jason Derenick, John Spletzer, and Brian Satterfield. "Little Ben: The Ben Franklin Racing Team's Entry in the 2007 DARPA Grand Challenge." *Journal of Field Robotics* 25, no. 9 (2008): 598–614.

Bojarski, Mariusz et al., "End to End Learning for Self-Driving Cars." *arXiv preprint arXiv:1604.07316* (2016), https://arxiv.org/abs/1604.07316.

Bollmer, Grant. "Books of Faces: Cultural Techniques of Basic Emotions." *NEC-SUS*, Spring (2019), https://necsus-ejms.org/books-of-faces-cultural-techniques-of-basic-emotions/.

Bradshaw, Tim. "How Apple's Ultra-wideband Chip Could Transform Its Products." *Financial Times*, September 20, 2019, www.ft.com/content/47e914a0-da3b-11e9-8f9b-77216ebe1f17.

Braid, Deborah, Alberto Broggi, and Gary Schmiedel. "The TerraMax Autonomous Vehicle." *Journal of Field Robotics* 23, no. 9 (2006): 693–708.

Brandom, Russell. "Microsoft Pulls Open Facial Recognition Dataset After *Financial Times* Investigation." *The Verge*, June 7, 2019, www.theverge.com/2019/6/7/18656800/microsoft-facial-recognition-dataset-removed-privacy.

Bratton, Benjamin H. *The Stack: On Software and Sovereignty*. Cambridge, MA: MIT Press, 2016.

Bridle, James. *New Dark Age: Technology and the End of the Future*. New York: Verso Books, 2018.

Brighenti, Andrea Mubi. *Visibility in Social Theory and Social Research*. Houndmills, Basingstoke, Hampshire: Palgrave Macmillan, 2010.

Broad, Ellen. *Made by Humans: The AI Condition*. Melbourne: Melbourne University Publishing, 2018.

Brown, Barry, and Eric Laurier. "The Trouble with Autopilots: Assisted and Autonomous Driving on the Social Road." *CHI 2017, May 6–11, 2017, Denver, CO*: 416–429.

Brown, C. Scott. "Google Updates Goggles After 3 Years Just to Tell People to Install Lens Instead." *Android Authority*, August 17, 2018, www.androidauthority.com/google-goggles-lens-896203/.

Brownlee, Jason. *Deep Learning for Computer Vision: Image Classification, Object Detection, and Face Recognition in Python*. Vermont, Victoria, Australia: Machine Learning Mastery, 2019.

Brumberger, Eva. "The Myth of Digital Literacy and Digital Natives." In *Digital Media: Transformations in Human Communication*, second edition. Edited by Paul Messaris and Lee Humphreys. New York: Peter Lang, 2017.

Bunz, Mercedes, and Graham Meikle. *The Internet of Things*. Cambridge: Polity Press, 2017.

Burgess, Jean, and Joshua Green. *YouTube: Online Video and Participatory Culture*, second edition. London: Polity Press, 2018.

Burnett, Ron. *How Images Think*. Cambridge, MA: MIT Press, 2004.

Busby, Martha. "People at King's Cross Site Express Unease About Facial Recognition." *The Guardian*, August 14, 2019, www.theguardian.com/technology/2019/aug/13/people-at-kings-cross-site-express-unease-about-facial-recognition.

Cai, Yongshun. "Grid Management and Social Control in China." *Asia Dialogue*, April 27, 2018, https://theasiadialogue.com/2018/04/27/grid-management-and-social-control-in-china/.

Cervantes, Edgar. "What is Night Mode and How Does it Work?" *Android Authority*, April 28, 2019, www.androidauthority.com/what-is-night-mode-and-how-does-it-work-979590/.

Chamayou, Grégoire. *Drone Theory*. London: Penguin, 2015.

Chan, Holmes. "Hong Kong Tech Firm Pulls Out of Smart Lamppost Programme after Surveillance Accusations and Staff Threats." *Hong Kong Free Press*, August 26, 2019, www.hongkongfp.com/2019/08/26/hong-kong-tech-firm-pulls-smart-lamppost-programme-surveillance-accusations-staff-threats/.

Chandler, Katherine. "American Kamikaze: Television-Guided Assault Drones in World War II." In *Life in the Age of Drone Warfare*. Edited by Lisa Parks and Karen Caplan, 89–111. Durham, NC: Duke University Press, 2017.

Chang, Haw-Shiuan, Erik Learned-Miller and Andrew McCallum. "Active Bias: Training More Accurate Neural Networks by Emphasizing High Variance Samples." In *Advances in Neural Information Processing Systems* (NIPS) 2017, *arXiv.org*, arXiv:1704.07433.

Chardon, Alain, Isabelle Cretois, and Colette Hourseau. "Skin Colour Typology and Suntanning Pathways." *International Journal of Cosmetic Science* 13, no. 4 (1991): 191–208.

Cheney-Lippold, John. *We Are Data: Algorithms and the Making of Our Digital Selves*. New York: NYU Press, 2018.

Chesher, Chris. "Between Image and Information: The iPhone Camera in the History of Photography." In *Studying Mobile Media: Cultural Technologies, Mobile Communication, and the iPhone*. Edited by Larissa Hjorth, Jean Burgess, and Ingrid Richardson, 98–117. New York: Routledge, 2012.

Chesney, Robert, and Danielle Citron. "Deepfakes and the New Disinformation War: The Coming Age of Post-Truth Geopolitics." *Foreign Affairs* 98, January/February (2019), www.foreignaffairs.com/articles/world/2018-12-11/deepfakes-and-new-disinformation-war.

Choi, Hanbyul, Jonghwa Park, and Yoonhyuk Jung. "The Role of Privacy Fatigue in Online Privacy Behavior." *Computers in Human Behavior* 81 (2018): 42–51.

"Civilian Drone Operated at Polish Riots 11–11–2011." YouTube video, 2:40. "disinpho," November 18, 2011, https://youtu.be/KOxh9dbkNT4.

Conger, Kate, Richard Fausset, and Serge F. Kovaleski. "San Francisco Bans Facial Recognition Technology." *New York Times*, May 14, 2019, www.nytimes.com/2019/05/14/us/facial-recognition-ban-san-francisco.html.

Cooley, Heidi Rae. "It's All About the *Fit*: The Hand, the Mobile Screenic Device and Tactile Vision." *Journal of Visual Culture* 3, no. 2 (2004): 133–155.

Crampton, Jeremy W., Mark Graham, Ate Poorthuis, Taylor Shelton, Monica Stephens, Matthew W. Wilson, and Matthew Zook. "Beyond the Geotag: Situating 'Big Data' and Leveraging the Potential of the Geoweb." *Cartographic and Geographic Information Science* 40, no. 2 (2013): 130–139.

Crary, Jonathan. *Techniques of the Observer: Vision and Modernity in the Nineteenth Century*. Cambridge, MA: MIT Press, 1990.

Crawford, Kate, and Trevor Paglen. "Excavating AI: The Politics of Images in Machine Learning Training Sets." *excavating.ai*, September 19, 2019, https://excavating.ai.

Cubitt, Sean. *The Practice of Light: A Genealogy of Visual Technologies from Prints to Pixels*. Cambridge, MA: MIT Press, 2014.

Datta, Soumya Kanti, Rui Pedro Ferreira Da Costa, Jérôme Härri, and Christian Bonnet. "Integrating Connected Vehicles in Internet of Things Ecosystems: Challenges and Solutions." In *Proceedings of the 2016 IEEE 17th International Symposium on A World of Wireless, Mobile and Multimedia Networks (WoWMoM)*, 1–6. New York: IEEE, 2016.

Debes, John L. "The Loom of Visual Literacy: An Overview." *Audiovisual Instruction* 14, no. 8 (1969): 25–27.

de Certeau, Michel. *The Practice of Everyday Life*. Translated by Steven Rendall. Berkeley, CA: University of California Press, 1984.

Deleuze, Gilles. *Cinema I: The Movement-Image*. Translated by Hugh Tomlinson and Barbara Habberjam. Minneapolis, MN: University of Minnesota Press, 1986.

Deleuze, Gilles. *Cinema II: The Time-Image*. Translated by Hugh Tomlinson and Barbara Habberjam. Minneapolis, MN: University of Minnesota Press, 1989.

Deleuze, Gilles, and Felix Guattari. *A Thousand Plateaus: Capitalism and Schizophrenia*. Translated by Brian Massumi. London: Continuum, 2002.

de Souza e Silva, Adriana, and Jordan Frith. *Net Locality: Why Location Matters in a Networked World*. London: Wiley, 2011.

Dewey, John. *Essays in Experimental Logic*. Chicago, IL: University of Chicago Press, 1916.

Dickmanns, Ernst D., Reinhold Behringer, Dirk Dickmanns, Tobias Hildebrandt, Markus Maurer, Frank Thomanek and Joachim Schiehlen. "The Seeing Passenger Car 'VaMoRs-P'." *Proceedings of the IEEE Intelligent Vehicles '94 Symposium*, 68–73. New York: IEEE, 1994, https://doi.org/10.1109/IVS.1994.639472.

D'Onfro, Jillian. "This Map Shows Which Cities Are Using Facial Recognition Technology – And Which Have Banned It." *Forbes*, July 18, 2019, www.forbes.com/sites/jilliandonfro/2019/07/18/map-of-facial-recognition-use-resistance-fight-for-the-future/#40de1d587e61.

Dredge, Stuart. "Five of the Best Face Swap Apps." *The Guardian*, March 17, 2016, www.theguardian.com/technology/2016/mar/17/five-of-the-best-face-swap-apps.

DroneZon. "12 Top Collision Avoidance Drones and Obstacle Detection Explained." *DroneZon*, December 28, 2018, www.dronezon.com/learn-about-drones-quadcopters/top-drones-with-obstacle-detection-collision-avoidance-sensors-explained/.

Ekman, Paul, and Wallace V. Friesen. *Unmasking the Face: A Guide to Recognizing Emotions from Facial Clues*. San Jose, CA: ISHK, 2003.

Eshet-Alkalai, Yoram. "Digital Literacy: A Conceptual Framework for Survival Skills in the Digital Era." *Journal of Educational Multimedia and Hypermedia* 13, no. 1 (2004): 93–106.

Eubanks, Virginia. *Automating Inequality: How High-tech Tools Profile, Police, and Punish the Poor*. New York: St. Martin's Press, 2018.

Eubanks, Virginia. "Trapped in the Digital Divide: The Distributive Paradigm in Community Informatics." *The Journal of Community Informatics* 3, no. 2 (2007), http://blog.ci-journal.net/index.php/ciej/article/view/293.

Eureka, William. "Simplify Before You Automate." *Material Handling & Logistics*, April 1, 2010, www.mhlnews.com/technology-amp-automation/simplify-you-automate#close-olyticsmodal.

Evans, Jake. "Driverless Cars: Kangaroos Throwing Off Animal Detection Software." *ABC News*, June 24, 2017, www.abc.net.au/news/2017-06-24/driverless-cars-in-australia-face-challenge-of-roo-problem/8574816.

Evans, Leighton, and Michael Saker. *Location-based Social Media: Space, Time and Identity*. Houndmills, Basingstoke: Palgrave Macmillan, 2017.

Evans, Sandra K., Katy E. Pearce, Jessica Vitak, and Jeffrey W. Treem. "Explicating Affordances: A Conceptual Framework for Understanding Affordances in Communication Research." *Journal of Computer-mediated Communication* 22, no. 1 (2017): 35–52.

Eye in the Sky. Directed by Gavin Hood. London: Entertainment One/Raindog Films, 2015.

"Face Recognition Homepage." www.face-rec.org/databases/.

Farkas, Leslie G. *Anthropometry of the Head and Face*. New York: Raven Press, 1994.

Farkas, Leslie G., Marko J. Katic, and Christopher R. Forrest. "International Anthropometric Study of Facial Morphology in Various Ethnic Groups/Races." *Journal of Craniofacial Surgery* 16, no. 4 (2005): 615–646.

Farman, Jason. *Mobile Interface Theory: Embodied Space and Locative Media*. New York: Routledge, 2012.

Feenberg, Andrew. "The Ambivalence of Technology." *Sociological Perspectives* 33, no. 1 (1990): 35–50.

Fight for the Future. "Ban Facial Recognition." www.banfacialrecognition.com/map/.

"Fight for the Future." www.fightforthefuture.org/.

"First Prime Air Delivery." *Amazon Prime Air*, www.amazon.com/Amazon-Prime-Air/b?ie=UTF8&node=8037720011.

Fischer, Frank. *Citizens, Experts and the Environment: The Politics of Local Knowledge*. Durham, NC: Duke University Press, 2000.

Floreano, Dario, and Robert J. Wood. "Science, Technology and the Future of Small Autonomous Drones." *Nature* 521, no. 7553 (2015): 460–466.

Flusser, Vilém. *Into the Universe of Technical Images*. Minneapolis, MN: University of Minnesota Press, 2011.

Flyability. "Elios – Inspect & Explore Indoor and Confined Spaces." *Flyability*, www.flyability.com/elios/.

Fort, Kären, Gilles Adda, and K. Bretonnel Cohen. "Amazon Mechanical Turk: Gold Mine or Coal Mine?" *Computational Linguistics* 37, no. 2 (2011): 413–420.

Foster, Hal, ed. *Vision and Visuality*. Seattle: Bay Press, 1988.

Fraedrich, Eva, and Barbara Lenz. "Societal and Individual Acceptance of Autonomous Driving." In *Autonomous Driving*. Edited by Markus Maurer, J. Christian Gerdes, Barbara Lenz and Hermann Winner, 621–640. Berlin: Springer, 2016.

Freemantle, Harry. *Seeing the Social: Selected Visibility Technologies*. Fremantle: Vivid Publishing, 2010.

Fried, Ina. "Tiny Chipmaker Movidius Has a Tiny Chip That Looms Large." *Recode*, March 16, 2016, www.recode.net/2016/3/16/11587020/tiny-chipmaker-movidius-has-a-tiny-chip-that-you-will-be-seeing-a-lot.

Frith, Jordan. *Smartphones as Locative Media*. Cambridge: Polity Press, 2015.

Futuramille. "First Person View Equipment." *Drone Racing Pilots Forums*, September 8, 2018, https://droneracingpilots.com/threads/kinda-bummed-flysight-spexman-2-goggles.2064/#post-18942.

Galal, Amr Mohamed. "An Analytical Study on the Modern History of Digital Photography." *International Design Journal* 6, no. 2 (2016): 203–215.

Garrett, Bradley L., and Anthony McCosker. "Non-Human Sensing: New Methodologies for the Drone Assemblage." In *Refiguring Techniques in Digital Visual Research*. Edited by Edgar Gómez Cruz, Shanti Sumartojo, and Sarah Pink, 13–23. Basingstoke: Palgrave Macmillan, 2017.

Gates, Kelly A. *Our Biometric Future: Facial Recognition Technology and the Culture of Surveillance*. New York: NYU Press, 2011.

Gehl, Robert W. "The Archive and the Processor: The Internal Logic of Web 2.0." *New Media & Society* 13, no. 8 (2011): 1228–1244.

Geospatial World. "Automatic Vertical Scanning for Drones Now Available." *Geospatial World*, February 22, 2019, www.geospatialworld.net/news/%D0%B0utomatic-vertical-scanning-for-drones-now-available/.

Ghaffary, Shirin, and Rani Molla. "Here's Where the US Government Is Using Facial Recognition Technology to Surveil Americans." *Vox/Recode*, July 18, 2019, www.vox.com/recode/2019/7/18/20698307/facial-recognition-technology-us-government-fight-for-the-future.

Gibson, James J. *The Ecological Approach to Visual Perception*. London: Lawrence Erlbaum, 1986.

Gillespie, Tarleton. *Custodians of the Internet: Platforms, Content Moderation, and the Hidden Decisions that Shape Social Media*. New Haven, CT: Yale University Press, 2018.

Gillespie, Tarleton, Pablo J. Boczkowski, and Kirsten A. Foot, eds. *Media Technologies: Essays on Communication, Materiality, and Society*. Cambridge, MA: MIT Press, 2013.

"The Giver of Names (1991–) by David Rokeby." YouTube video, 5:59. "David Rokeby," December 8, 2006, https://youtu.be/sO9RggYz24Q.

Glister, Paul. *Digital Literacy*. New York: John Wiley & Sons, 1997.

Goggin, Gerard. *Cell Phone Culture: Mobile Technology in Everyday Life*. New York: Routledge, 2006.

Goggin, Gerard. *Global Mobile Media*. London: Routledge, 2011.

Gómez Cruz, Edgar, and Helen Thornham. "Selfies Beyond Self-representation: The (Theoretical) F(r)ictions of a Practice." *Journal of Aesthetics & Culture* 7, no. 1 (2015), https://doi.org/10.3402/jac.v7.28073.

Goodfellow, Ian J., Jean Pouget-Abadie, Mehdi Mirza, Bing Xu, David Warde-Farley, Sherji Ozair, Aaron Courville, Yoshua Bengio. "Generative Adversarial Nets." In *Proceedings in Advances in Neural Information Processing Systems 27 (NIPS 2014)*: 2672–2680.

Gopro, Inc. "Image Sensor Data Compression and DSP Decompression." *US Patent Office*, US20160227068 A1, August 4, 2016, www.google.com/patents/US20160227068.

Graham, Connor, Eric Laurier, Vincent O'Brien, and Mark Rouncefield. "New Visual Technologies: Shifting Boundaries, Shared Moments." *Visual Studies* 26, no. 2 (2011): 87–91.

Grusin, Richard, ed. *The Nonhuman Turn*. Minneapolis, MN: University of Minnesota Press, 2015.

Gustuvson, Todd. *Camera: A History of Photography from Daguerreotype to Digital*. New York: Sterling, 2009.

Gye, Lisa. "Picture This: The Impact of Mobile Camera Phones on Personal Photographic Practices." In *Mobile Phone Cultures*. Edited by Gerard Goggin, 135–144. New York: Routledge, 2008.

Haggerty, Kevin D., and Richard V. Ericson. "The Surveillant Assemblage." *The British Journal of Sociology* 51, no. 4 (2000): 605–622.

Halpern, Megan, and Lee Humphreys. "iPhoneography as an Emergent Art World." *New Media & Society* 18, no. 1 (2016): 62–81.

Hand, Martin. *Ubiquitous Photography*. Cambridge: Polity Press, 2012.

Haraway, Donna. *Simians, Cyborgs and Women: The Reinvention of Nature.* New York: Routledge, 1991.

Hawkins, Andrew J. "Tesla Didn't Fix an Autopilot Problem for Three Years, and Now Another Person is Dead." *The Verge,* May 17, 2019, www.theverge.com/2019/5/17/18629214/tesla-autopilot-crash-death-josh-brown-jeremy-banner.

Hayes, Brian. "Computational Photography." *American Scientist* 96, no. 2 (2008), www.americanscientist.org/article/computational-photography.

Helmond, Anne. "The Platformization of the Web: Making Web Data Platform Ready." *Social Media + Society* July–December (2015): 1–11, https://doi.org/10.1177/2056305115603080.

Hess, Aaron. "Selfies| The Selfie Assemblage." *International Journal of Communication* 9 (2015): 1629–1646, https://ijoc.org/index.php/ijoc/article/view/3147/1389.

Hjorth, Larissa, Jean Burgess, and Ingrid Richardson, eds. *Studying Mobile Media: Cultural Technologies, Mobile Communication, and the iPhone.* New York: Routledge, 2012.

Hjorth, Larissa, and Sarah Pink. "New Visualities and the Digital Wayfarer: Reconceptualizing Camera Phone Photography and Locative Media." *Mobile Media and Communication* 2 no. 1 (2014): 40–57.

Hochman, Nadav. "The Social Image." *Big Data & Society,* July–September (2014): 1–15.

Hoffman, Samantha. "Social Credit." *Australian Strategic Policy Institute,* June 28, 2018, www.aspi.org.au/report/social-credit.

Horning, Rob. "Selfies Without the Self." *The New Inquiry,* November 23, 2014, http://thenewinquiry.com/blogs/marginal-utility/selfies-without-the-self/.

Humphreys, Lee. *The Qualified Self: Social Media and the Accounting of Everyday Life.* Cambridge, MA: MIT Press, 2018.

Hutchby, Ian. "Technologies, Texts and Affordances," *Sociology* 3, no. 2 (2001): 444, 441–456.

Ingold, Tim. *Lines: A Brief History.* London: Routledge, 2007.

Isin, Engin, and Evelyn Ruppert. *Being Digital Citizens.* London: Rowman and Littlefield International, 2015.

Jablonowski, Maximilian. "Would You Mind My Drone Taking a Picture of Us?" *Photomediations Machine,* September 29, 2014, http://photomediationsmachine.net/2014/09/29/would-you-mind-my-drone-taking-a-picture-of-us/.

Jacknis, Ira. "Margaret Mead and Gregory Bateson in Bali: Their Use of Photography and Film." *Cultural Anthropology* 3, no. 2 (1988): 160–177.

Jafri, Rabia, Rodrigo Louzada Campos, Syed Abid Ali, and Hamid R. Arabnia. "Utilizing the Google Project Tango Tablet Development Kit and the Unity Engine for Image and Infrared Data-Based Obstacle Detection for the Visually Impaired." In *International Conference on Health Informatics and Medical Systems | HIMS'16,* 163–164 (2016), https://pdfs.semanticscholar.org/6572/49b121859df27bd8b0678767e9df4335663b.pdf.

Jafri, Rabia, and Marwa Mahmound Khan. "Obstacle Detection and Avoidance for the Visually Impaired in Indoor Environments Using Google's Project Tango Device." In *Computers Helping People with Special Needs. ICCHP 2016. Lecture Notes in Computer Science, vol 9759.* Edited by K. Miesenberger, C. Bühler, and P. Penaz, 179–185. Cham, Switzerland: Springer, 2016, https://doi.org/10.1007/978-3-319-41267-2_24.

James, William. *Essays in Radical Empiricism.* New York: Longmans Green and Co., 1912.

Jenkins, Reese V. "Technology and the Market: George Eastman and the Origins of Mass Amateur Photography." *Technology and Culture* 16, no. 1 (1975): 1–19.

Jing, Meng. "From Travel and Retail to Banking, China's Facial-recognition Systems Are Becoming Part of Daily Life." *South China Morning Post*, February 8, 2018, www.scmp.com/tech/social-gadgets/article/2132465/travel-and-retail-banking-chinas-facial-recognition-systems-are.

Johnston, John. "Machinic Vision." *Critical Inquiry* 26 no. 1 (1999): 27–48.

Jones, Rodney H., and Christoph A. Hafner. *Understanding Digital Literacies: A Practical Introduction*. New York: Routledge, 2012.

Jurgenson, Nathan. *The Social Photo: On Photography and Social Media*. New York: Verso, 2019.

Kato, Fumitoshi. "Seeing the Seeing of Others: Conducting a Field Study with Mobile Phones/Camera Phones." In *Proceedings of the conference Seeing, Understanding, Learning in the Mobile Age, Budapest, June 10–12, 2004*.

Kato, Fumitoshi, Daisuke Okabe, Mizuko Ito, and Ryuhei Uemoto. "Uses and Possibilities of the Keitai Camera." In *Personal, Portable, Pedestrian, Mobile Phones in Japanese Life*. Edited by Mizuko Ito, Misa Matsuda, and Daisuke Okabe, 300–310. Cambridge, MA: MIT Press, 2005.

Katz, James E., and Mark A. Aakhus, eds. *Perpetual Contact: Mobile Communication, Private Talk, Public Performance*. Cambridge: Cambridge University Press, 2002.

Kearney, Alex, and Oliver Oxton. "Making Meaning: Semiotics Within Predictive Knowledge Architectures." *arXiv.org, arXiv:1904.09023*.

Kember, Sarah, and Joanna Zylinska. *Life After New Media: Mediation as a Vital Process*. Cambridge, MA: MIT Press, 2012.

Khosla, Aditya, Tinghui Zhou, Tomasz Malisiewicz, Alexei A. Efros, and Antonio Torralba. "Undoing the Damage of Dataset Bias." In *European Conference on Computer Vision*, 158–171. Berlin: Springer, 2012.

Kjoelen, Arve, M. J. Thompson, Scott E. Umbaugh, Randy Hays, and William V. Stoecker. "Performance of AI Methods in Detecting Melanoma." *IEEE Engineering in Medicine and Biology Magazine* 14, no. 4 (1995): 411–416.

Kleinman, Zoe. "King's Cross Developer Defends Use of Facial Recognition." *BBC News*, August 12, 2019, www.bbc.com/news/technology-49320520.

Knobel, Michele, and Colin Lankshear, eds. *Digital Literacies: Concepts, Policies and Practices*. New York: Peter Lang, 2008.

Koskinen, Ilpo. "Seeing with Mobile Images: Towards the Perpetual Visual Contact." In *Proceedings of the conference The Global and the Local in Communication: Places, Images, People, Connections, Budapest, June 10–12, 2004*.

Kurgan, Laura. *Close Up at a Distance: Mapping, Technology and Politics*. New York: Zone Books, 2016.

Lankshear, Colin, and Michele Knobel. "Introduction: Digital Literacies— Concepts, Policies and Practices." In *Digital Literacies Concepts, Policies and Practices*. Edited by M. Knobel and C. Lankshear, 1–16, 4. New York: Peter Lang, 2008.

Lasén, Amparo. "Digital Self-Portraits, Exposure and the Modulation of Intimacy." In *Mobile and Digital Communication: Approaches to Public and Private*. Edited by J.R. Carvelheiro and A.S. Telleria. Covilhã, Portugal: Livros LabCom, 2015.

Lasén, Amparo, and Edgar Gómez-Cruz. "Digital Photography and Picture Sharing: Redefining the Public/Private Divide." *Knowledge, Technology and Policy*, no. 22 (2009): 205–215.

Lee, Jaihyun. "Optimization of a Modular Drone Delivery System." In *Proceedings of 2017 Annual IEEE International Systems Conference (SysCon)*, 1–8. New York: IEEE, 2017.

Leskin, Paige. "Since Going Viral Again for Making People Look Old, FaceApp Has Been Downloaded by 12.7 Million New Users." *Business Insider*, July 19, 2019, www.businessinsider.com.au/faceapp-viral-downloads-13-million-new-users-last-week-2019-7?r=US&IR=T.

Leszczynski, Agnieszka. "Situating the Geoweb in Political Economy." *Progress in Human Geography* 36, no. 1 (2012): 72–89.

Liao, Tony, and Lee Humphreys. "Layar-ed Places: Using Mobile Augmented Reality to Tactically Reengage, Reproduce, and Reappropriate Public Space." *New Media & Society* 17, no. 9 (2015): 1418–1435.

Lim, Ye Seul, Phu Hien La, Jong Soo Park, Mi Hee Lee, Mu Wook Pyeon, and Jee-In Kim. "Calculation of Tree Height and Canopy Crown From Drone Images Using Segmentation." *Journal of the Korean Society of Surveying, Geodesy, Photogrammetry and Cartography* 33, no. 6 (2015): 605–613.

Limer, Eric. "Microsoft's New Real-World Search Engine is Incredible and Horrifying." *Popular Mechanics*, May 10, 2017, www.popularmechanics.com/technology/infrastructure/news/a26456/microsoft-machine-learning-vision/.

Ling, Rich. *The Mobile Connection: The Cell Phone's Impact on Society*. San Francisco, CA: Morgan Kaufmann, 2004.

Ling, Rich. *New Tech, New Ties: How Mobile Communication Is Reshaping Social Cohesion*. Cambridge, MA: MIT Press, 2008.

Lister, Martin. "Is the Camera an Extension of the Photographer?" In *Digital Photography and Everyday Life*. Edited by Edgar Gómez Cruz and Asko Lehmuskallio, 209–214. New York: Routledge, 2016.

Little, Anthony C., Benedict C. Jones, and Lisa M. DeBruine. "Facial Attractiveness: Evolutionary Based Research." *Philosophical Transactions of the Royal Society* 366 (2011): 1638–1659.

Liu, Yanxi, Karen L. Schmidt, Jeffrey F. Cohn, and Sinjini Mitra. "Facial Asymmetry Quantification for Expression Invariant Human Identification." *Computer Vision and Image Understanding* 91, no. 1–2 (2003): 138–159.

Liu, Ziwei, Ping Luo, Xiaogang Wang, and Xiaoou Tang. "Deep Learning Face Attributes in the Wild." *International Conference on Computer Vision (ICCV)*, 3730–3738. New York: IEEE, 2015.

Lobinger, Katharina. "Photographs as Things – Photographs of Things. A Texto-material Perspective on Photo-sharing Practices." *Information, Communication & Society* 19, no. 4 (2016): 475–488.

Lu, Ning, Nan Cheng, Ning Zhang, Xuemin Shen, and Jon W. Mark. "Connected Vehicles: Solutions and Challenges." *IEEE Internet of Things Journal* 1, no. 4 (2014): 289–299.

Luminar Technology. www.luminartech.com/technology/index.html.

Lyon, David. *Surveillance After Snowden*. London: Polity, 2015.

Mackenzie, Adrian. *Wirelessness: Radical Empiricism in Network Cultures*. Cambridge, MA: MIT Press, 2010.

Malo, Jim. "All of Australia's 15.2 Million Buildings Have Been Mapped." *The Age*, October 30, 2018, https://bit.ly/2qoo9DT.

Mangalindan, J. P. "Uber Acquires Mapping Startup deCarta." *Mashable Australia*, March 4, 2015, http://mashable.com/2015/03/03/uber-acquires-mapping-decarta/#_hunXWPpFaqF.

Marchetti, Michael A., Noel C. F. Codella, Stephen W. Dusza, David A. Gutman, Brian Helba, Aadi Kalloo, Nabin Mishra, Cristina Carrera, M. Emre Celebri, Jennifer

L. DeFazio, Natalia Jaimes, Ashfaq A. Marghoob, Elizabeth Quigley, Alon Scope, Oriol Yélamos, and Allan C. Halpern. "Results of the 2016 International Skin Imaging Collaboration International Symposium on Biomedical Imaging Challenge: Comparison of the Accuracy of Computer Algorithms to Dermatologists for the Diagnosis of Melanoma from Dermoscopic Images." *Journal of the American Academy of Dermatology* 78, no. 2 (2018): 270–277.

Marshall, Aarian. "Why Self-driving Cars *Can't Even* With Construction Zones." *Wired*, February 10, 2017, www.wired.com/2017/02/self-driving-cars-cant-even-construction-zones/?mbid=social_gplus.

Marvin, Carolyn. "Your Smart Phones Are Hot Pockets to Us: Context Collapse in a Mobilized Age." *Mobile Media & Communication* 1, no. 1 (2013): 153–159.

Massumi, Brian. "The Supernormal Animal." In *The Nonhuman Turn*. Edited by Richard Grusin, 1–17. Minneapolis, MN: University of Minnesota Press, 2015.

Mayer-Schönberger, Viktor. *Delete: The Virtue of Forgetting in the Digital Age*. Princeton, NJ: Princeton University Press, 2011.

McCarthy, John, and Patrick J. Hayes. "Some Philosophical Problems from the Standpoint of Artificial Intelligence." In *Readings in Artificial Intelligence*. Edited by Bonnie Lynn Webber and Nils J. Nilsson, 431–450. Burlington, MA: Morgan Kaufman, 1981.

McCosker, Anthony. "Drone Media: Unruly Systems, Radical Empiricism and Camera Consciousness." *Culture Machine* 16 (2015).

McCosker, Anthony. "Drone Vision, Zones of Protest, and the New Camera Consciousness." *Media Fields* 9 (2015).

McCosker, Anthony, Son Vivienne, and Amelia Johns, eds. *Negotiating Digital Citizenship: Control, Contest and Culture*. London: Rowman & Littlefield International, 2016.

McStay, Andrew. *Emotional AI: The Rise of Empathic Media*. London: Sage, 2018.

Merchant, Brian. *The One Device: The Secret History of the iPhone*. London: Bantam Press, 2017.

Merlan, Anna, and Dhruv Mehrotra. "Amazon's Facial Analysis Program Is Building a Dystopic Future for Trans and Nonbinary People." *Jezebel*, June 27, 2019, https://jezebel.com/amazons-facial-analysis-program-is-building-a-dystopic-1835075450.

Merleau-Ponty, Maurice. *Phenomenology of Perception*. Translated by C. Smith. New York: Routledge, 1962 [1945].

Miller, Daniel, and Jolynna Sinanan. *Webcam*. New York: John Wiley & Sons, 2014.

Mitchell, William J. T. *Picture Theory: Essays on Verbal and Visual Representation*. Chicago, IL: University of Chicago Press, 2007 [1994].

Mitchell, William J. *The Reconfigured Eye: Visual Truth in the Post-Photographic Era*. Cambridge, MA: MIT Press, 1992.

Mittelstadt, Brent. "Automation, Algorithms, and Politics| Auditing for Transparency in Content Personalization Systems." *International Journal of Communication* 10 (2016): 4991–5002, https://ijoc.org/index.php/ijoc/article/view/6267/1808.

Mlot, Stephanie. "New Systems Teach Driverless Cars to 'See'." *PC Mag*, December 24, 2015, http://au.pcmag.com/software/40861/news/new-systems-teach-driverless-cars-to-see.

Mochizuki, Saku, Jun Kataoka, Leo Tagawa, Yasuhiro Iwamoto, Hiroshi Okochi, Naoya Katsumi, Shuntaro Kinno, Makoto Arimoto, Takuya Maruhashi, Kazuya Fujieda, Takuya Kurihara, and Shinji Ohsuka. "First Demonstration of Aerial Gamma-ray Imaging Using Drone for Prompt Radiation Survey in Fukushima." *Journal of Instrumentation* 12, November (2017), https://doi.org/10.1088/1748-0221/12/11/P11014.

Montemerlo, Michael, et al. "Junior: The Stanford Entry in the Urban Challenge." *Journal of Field Robotics* 25, no. 9 (2008): 569–597.

Moulard-Leonard, Valentine. *Bergson-Deleuze Encounters: Transcendental Experience and the Thought of the Virtual.* Albany, NY: State University of New York Press, 2008.

"Movidius and DJI Bring Vision-Based Autonomy to Phantom 4." YouTube video, 3:28. "Intel Movidius," March 16, 2016, https://youtu.be/hX0UELNRR1I?list= PL0PjXgwbk-vur0wGByQSVu9r1ODn5TK8m.

Mozur, Paul. "Inside China's Dystopian Dreams: A.I., Shame and Lots of Cameras." *New York Times*, October 7, 2018, www.nytimes.com/2018/07/08/business/china-surveillance-technology.html.

Mozur, Paul. "One Month, 500,000 Face Scans: How China Is Using A.I. to Profile a Minority." *New York Times*, April 14, 2019, www.nytimes.com/2019/04/14/tech nology/china-surveillance-artificial-intelligence-racial-profiling.html.

Mueller, Robert. "Google Adds New Tricks and Devices to Tango Augmented Reality." *Fast Company*, May 17, 2017, https://news.fastcompany.com/google-adds-new-tricks-and-devices-to-tango-augmented-reality-4038110.

Munster, Anna. *An Aesthesia of Networks: Conjunctive Experience in Art and Technology.* Cambridge, MA: MIT Press, 2013.

Munster, Anna. "Transmateriality: Toward an Energetics of Signal in Contemporary Mediatic Assemblages." *Cultural Studies Review* 20 no. 1 (2014): 150–167.

Murray, Susan. "Digital Images, Photo-Sharing, and Our Shifting Notions of Everyday Aesthetics." *Journal of Visual Culture* 7, no. 2 (2008): 147–163.

Ninjajimmy83. "Getting started in FPV – Advice or recommendations." *Reddit, r/fpvracing*, April 22, 2019, www.reddit.com/r/fpvracing/comments/bfsgwb/getting_started_in_fpv_advice_or_recommendations/.

Noble, Safiya Umoja. *Algorithms of Oppression: How Search Engines Reinforce Racism.* New York: New York University Press, 2018.

Oakden-Rayner, Luke. "Exploring the ChestXray14 Dataset: Problems." *Luke Oakden-Rayner* (blog). December 18, 2017, https://lukeoakdenrayner.wordpress.com/2017/12/18/the-chestxray14-dataset-problems/.

Oh, Je-Keun, Giho Jang, Semin Oh, Jeong Ho Lee, Byung-Ju Yi, Young Shik Moon, John Seh Lee, and Youngjin Choi. "Bridge Inspection Robot System with Machine Vision." *Automation in Construction* 18, no. 7 (2009): 929–941.

O'Neil, Luke. "Doctored Video of Sinister Mark Zuckerberg Puts Facebook to the Test." *The Guardian*, 12 June, 2019, www.theguardian.com/technology/2019/jun/11/deepfake-zuckerberg-instagram-facebook.

Özkan, Türker, Timo Lajunen, Joannes El. Chliaoutakis, Dianne Parker, and Heikki Summala. "Cross-cultural Differences in Driving Behaviours: A Comparison of Six Countries." *Transportation Research Part F: Traffic Psychology and Behaviour* 9, no. 3 (2006): 227–242.

Packer, Jeremy, and Stephen B. Crofts Wiley, eds. *Communication Matters: Materialist Approaches to Media, Mobility and Networks.* London: Routledge, 2012.

Paglen, Trevor. "Invisible Images (Your Pictures Are Looking at You)." *The New Inquiry*, December 8, 2016, https://thenewinquiry.com/invisible-images-your-pictures-are-looking-at-you/.

Palmer, Daniel. "iPhone Photography: Mediating Visions of Social Space." In *Studying Mobile Media: Cultural Technologies, Mobile Communication, and the iPhone.* Edited by Larissa Hjorth, Jean Burgess, and Ingrid Richardson, 85–97. New York: Routledge, 2012.

Palmer, Daniel. "Photography as Indexical Data: Hans Eijkelboom and Pattern Recognition Algorithms." In *Photography and Ontology: Unsettling Images*. Edited by Donna West Brett and Natalya Lusty, 125–139. New York: Routledge, 2019.

Palmer, Daniel. "Redundant Photographs: Cameras, Software and Human Obsolescence." In *On the Verge of Photography*. Edited by Daniel Rubenstein, Johnny Goldin, and Andy Fisher, 49–67. Birmingham: ARTicle Press, 2013.

Palmer, Daniel. "The Rhetoric of the JPEG." In *The Photographic Image in Digital Culture*, second edition. Edited by Martin Lister, 149–164. London: Routledge, 2013.

Parikka, Jussi. *Digital Contagions: A Media Archaeology of Computer Viruses*. New York: Peter Lang, 2007.

Parikka, Jussi. *Insect Media: An Archaeology of Animals and Technology*. Minneapolis, MN: University of Minnesota Press, 2010.

Paris, Britt, and Joan Donovan. *Deepfakes and Cheap Fakes: The Manipulation of Visual Evidence*. New York: Data & Society, 2019, https://datasociety.net/wp-content/uploads/2019/09/DS_Deepfakes_Cheap_FakesFinal-1.pdf.

Parks, Lisa. *Cultures in Orbit: Satellites and the Televisual*. Durham, NC: Duke University Press, 2005.

Parks, Lisa. "Mapping Orbit: Toward a Vertical Public Sphere." In *Public Space, Media Space*. Edited by Chris Berry, Janet Harbord, and Rachel Moore, 61–87. Houndmills, Basingstoke: Palgrave Macmillan, 2013.

Parks, Lisa. "Vertical Mediation and the U.S. Drone War in the Horn of Africa." In *Life in the Age of Drone Warfare*. Edited by Lisa Parks and Karen Caplan, 134–157. Durham, NC: Duke University Press, 2017.

Parks, Lisa, and Karen Caplan, eds. *Life in the Age of Drone Warfare*. Durham, NC: Duke University Press, 2017.

Patel, Nilay. "How Google Controls Android: Digging Deep into the Skyhook Filings." *The Verge*, May 12, 2011, www.theverge.com/2011/05/12/google-android-skyhook-lawsuit-motorola-samsung.

Paul, Christiane. "Renderings of Digital Art." *Leonardo* 35, no. 5 (2002): 471–474, 476–484.

Peer, Eyal, Joachim Vosgerau, and Alessandro Acquisti. "Reputation as a Sufficient Condition for Data Quality on Amazon Mechanical Turk." *Behavior Research Methods* 46, no. 4 (2014): 1023–1031.

Peters, Chris, and Stuart Allan. "Everyday Imagery: Users' Reflections on Smartphone Cameras and Communication." *Convergence: The International Journal of Research into New Media Technologies* 24, no. 4 (2018): 357–373.

Peters, Jay. "Watch DARPA Test Out a Swarm of Drones." *The Verge*, August 9, 2019, www.theverge.com/2019/8/9/20799148/darpa-drones-robots-swarm-military-test.

Petrovskaya, Anna, and Sebastian Thrun, "Model Based Vehicle Detection and Tracking for Autonomous Urban Driving." *Autonomous Robots* 26 (2009): 123–130.

Pillai, Anil Narendran, "A Brief Introduction to Photogrammetry and Remote Sensing." *GIS Lounge*, July 12, 2015, www.gislounge.com/a-brief-introduction-to-photogrammetry-and-remote-sensing/.

Pinch, Trevor J., and Wiebe E. Bijker. "The Social Construction of Facts and Artefacts: Or How the Sociology of Science and the Sociology of Technology Might Benefit Each Other." *Social Studies of Science* 14, no. 3 (1984): 399–441.

Pinheiro, Pedro O., Ronan Collobert, and Piotr Dollár. "Learning to Segment Object Candidates." In *Proceedings of the 28th International Conference on Neural Information Processing Systems – Volume 2*, 1990–1998. New York: ACM, 2015.

Pinheiro, Pedro O., Tsung-Yi Lin, Ronan Collobert, and Piotr Dollár. "Learning to Refine Object Segments." In *European Conference on Computer Vision*, 75–91. Cham, Switzerland: Springer, 2016.

Pisters, Patricia. *The Matrix of Visual Culture: Working with Deleuze in Film Theory*. Stanford, CA: Stanford University Press, 2003.

"Planetary Skin Institute – Global Nervous System." YouTube video, 3:43. "Kostas," November 28, 2010, https://youtu.be/K3K990kVLS0.

Plantin, Jean-Christophe, Carl Lagoze, Paul N. Edwards, and Christian Sandvig. "Infrastructure Studies Meet Platform Studies in the Age of Google and Facebook." *New Media & Society* 20, no. 1 (2018): 293–310.

Poepolov, Dimtry A. "Semiotic Models in Artificial Intelligence Problems." In *IJCAI '75 Proceedings of the 4th International Joint Conference on Artificial Intelligence – Volume 1*, 65–70. New York: ACM, 1975, www.ijcai.org/Proceedings/75/Papers/010.pdf.

"Police Shoot Down RC Quadrocopter in Turkey – Truthloader." YouTube video, 1:45. "Point," June 13, 2013, www.youtube.com/watch?v=_A-ufp5gY3s&t=6s.

Pomerantz, Jeffrey. *Metadata*. Cambridge, MA: MIT Press, 2015.

Porcheron, Aurélie, Emmanuelle Mauger, and Richard Russell. "Aspects of Facial Contrast Decrease with Age and Are Cues for Age Perception." *PLoS One* 8, no. 3 (2013), https://doi.org/10.1371/journal.pone.0057985.

Posters, Bill. "Gallery: 'Spectre' Launches (Press Release)." *Bill Posters*, May 29, 2019, http://billposters.ch/spectre-launch/.

Prioleau, Mark. "Intel Pays $15B for Mobileye: A Strategic Play for Data." *Medium*, March 14, 2017, https://medium.com/@mprioleau/intel-pays-15b-for-mobileye-a-strategic-play-for-data-672eacd8bb7a.

Provost, Foster. "Machine Learning from Imbalanced Data Sets 101." In *Proceedings of the AAAI'2000 Workshop on Imbalanced Data Sets* 68, no. 2000, 1–3. New York: AAAI Press, 2000.

Quain, John R. "What Self-driving Cars See." *New York Times*, May 25, 2017, www.nytimes.com/2017/05/25/automobiles/wheels/lidar-self-driving-cars.html.

Raji, Inioluwa Deborah, and Joy Buolamwini. "Actionable Auditing: Investigating the Impact of Publicly Naming Biased Performance Results of Commercial AI Products." In *Proceedings of the 2019 AAAI/ACM Conference on AI, Ethics, and Society (AIES-19)*. New York: ACM, 2019, https://dam-prod.media.mit.edu/x/2019/01/24/AIES-19_paper_223.pdf.

Ramanathan, Narayanan, and Rama Chellappa. "Modeling Age Progression in Young Faces," *International Conference on Computer Vision and Pattern Recognition (CVPR)*, 387–394. New York: IEEE, 2006, https://doi.org/10.1109/CVPR.2006.187.

Ratner, Alexander, Christopher De Sa, Sen Wu, Daniel Selsam and Christopher Ré. "Data Programming: Creating Large Training Sets, Quickly." In *30th Conference on Neural Information Processing Systems (NIPS 2016), Barcelona, Spain*, https://papers.nips.cc/paper/6523-data-programming-creating-large-training-sets-quickly.pdf.

Richardson, Ingrid, and Rowan Wilken. "Parerga of the Third Screen: Mobile Media, Place, and Presence." In *Mobile Technology and Place*. Edited by Rowan Wilken and Gerard Goggin, 181–197. New York: Routledge, 2012.

Ritchin, Fred. *After Photography*. New York: WW Norton & Company, 2009.

Rosenfeld, Azriel. *Picture Processing by Computer*. New York: Academic Press, 1969.

Rothe, Rasmus, Radu Timofte, and Luc Van Gool. "Deep Expectation of Real and Apparent Age from a Single Image Without Facial Landmarks." *International Journal of Computer Vision* 126, no. 2–4 (2018): 144–157.

Rothstein, Adam. *Drone*. New York: Bloomsbury Academic, 2015.

Rubenstein, Daniel, and Katrina Sluis. "A Life More Photographic: Mapping the Networked Image." *Photographies* 1, no. 1 (2008): 9–28.

Sandvig, Christian, Kevin Hamilton, Karrie Karahalios, and Cedric Langbort. "An Algorithm Audit." In *Data and Discrimination: Collected Essaysi*. Edited by Seeta Peña Gangadharan, Virginia Eubanks, and Solon Barocas, 6–10. New York: New America, Open Technology Institute, 2014.

Sato, Yuki, Shingo Ozawa, Yuta Terasaka, Masaaki Kaburagi, Yuta Tanifuji, Kuniaki Kawabata, Hiroko Nakamura Miyamura, Ryo Izumi, Toshikazu Suzuki, and Tasuo Torii. "Remote Radiation Imaging System Using a Compact Gamma-ray Imager Mounted on a Multicopter Drone." *Journal of Nuclear Science and Technology* 55, no. 1 (2018): 90–96.

Schlegl, Thomas, Philipp Seeböck, Sebastian M. Waldstein, Ursula Schmidt-Erfurth, and Georg Langs. "Unsupervised Anomaly Detection with Generative Adversarial Networks to Guide Marker Discovery." In *Information Processing in Medical Imaging. IPMI 2017. Lecture Notes in Computer Science, vol 10265*. Edited by Marc Niethammer, Martin Styner, Stephen Aylward, Hongtu Zhu, Ipek Oguz, Pew-Thian Yap, and Dinggang Shen, 146–157. Cham, Switzerland: Springer, 2017.

Schwab, Klaus. "The Fourth Industrial Revolution: What it Means, How to Respond." *World Economic Forum*, January 14, 2016, www.weforum.org/agenda/2016/01/the-fourth-industrial-revolution-what-it-means-and-how-to-respond/.

Seaver, Nick. "Algorithms as Culture: Some Tactics for the Ethnography of Algorithmic Systems." *Big Data & Society* 4, no. 2 (2017), https://doi.org/10.1177/2053951717738104.

Senft, Theresa M. *Camgirls: Celebrity and Community in the Age of Social Networks*. New York: Peter Lang, 2008.

Senft, Theresa M., and Nancy K. Baym. "Selfies Introduction ~ What Does the Selfie Say? Investigating a Global Phenomenon." *International Journal of Communication* 9 (2015): 1588–1606, https://ijoc.org/index.php/ijoc/article/view/4067/1387.

Shaw, Spencer. *Film Consciousness: From Phenomenology to Deleuze*. Jefferson, NC: McFarland, 2008.

Shead, Sam. "BMW Hopes Google's Augmented Reality Tango Technology Will Help It Sell Cars." *Business Insider*, January 5, 2017, www.businessinsider.com/bmw-google-augmented-reality-tango-2017-1?r=US&IR=T&IR=T.

Simondon, Gilbert. *Du Mode d'existence des Objets Techniques*. Paris: Aubier, 1958.

Singer, Natasha. "Amazon Faces Investor Pressure Over Facial Recognition." *New York Times*, May 20, 2019, www.nytimes.com/2019/05/20/technology/amazon-facial-recognition.html.

Sklar, Max, Blake Shaw and Andrew Hogue. "Recommending Interesting Events in Real-time with Foursquare Check-ins." In *Proceedings of the sixth ACM conference on Recommender systems, September 9–13, 2012, Dublin, Ireland*, www.metablake.com/foursquare/recsys2012.pdf.

Smith, John R., with Joy Buolamwini and Timnit Gebru. "IBM Research Releases 'Diversity in Faces' Dataset to Advance Study of Fairness in Facial Recognition Systems." *IBM Blog*, January 29, 2019, www.ibm.com/blogs/research/2019/01/diversity-in-faces/.

Snickars, Pelle, and Patrick Vonderau, eds. *Moving Data: The iPhone and the Future of Media*. New York: Columbia University Press, 2012.

Solem, Jan Erik. *Programming Computer Vision with Python: Tools and Algorithms for Analyzing Images*. New York: O'Reilly Media, Inc., 2012.

Solon, Olivia. "Facial Recognition's 'Dirty Little Secret': Millions of Online Photos Scraped Without Consent." *NBC News*, March 12, 2019, www.nbcnews.com/tech/internet/facial-recognition-s-dirty-little-secret-millions-online-photos-scraped-n981921.

Sorokin, Alexander, and David Forsyth. "Utility Data Annotation with Amazon Mechanical Turk." In *2008 IEEE Computer Society Conference on Computer Vision and Pattern Recognition Workshops*, 1–8. New York: IEEE, 2008.

Souppouris, Aaron. "The German Car Industry is Buying Nokia's Here Maps." *Engadget*, August 3, 2015, www.engadget.com/2015/08/03/nokia-here-maps-sale-bmw-mercedes-audi/.

"Spridningstillstånd." *Lantmäteriet*, www.lantmateriet.se/sv/Om-Lantmateriet/Ratts information/spridningstillstand/.

Stafford, Barbara Maria. *Voyage into Substance: Art, Science, Nature, and the Illustrated Travel Account, 1760–1840*. Cambridge, MA: MIT Press, 1984.

Stark, Harold. "Introducing FaceApp: The Year of the Weird Selfie." *Forbes*, April 25, 2017, www.forbes.com/sites/haroldstark/2017/04/25/introducing-faceapp-the-year-of-the-weird-selfies/#334634e243d2.

Stewart, Jack. "Why Self-Driving Cars Need Superhuman Senses." *Wired*, September 11, 2017, www.wired.com/story/why-self-driving-cars-need-superhuman-senses/.

Stilgoe, Jack. "Machine Learning, Social Learning and the Governance of Self-driving Cars." *Social Studies of Science* 48, no. 1 (2018): 25–56.

"STUNNING IMAGES: Drone Films Poland Protest Pictures." YouTube video, 1:38. "On Demand News," November 16, 2011, https://youtu.be/gPLh8vkMZms.

Sullivan, Mark. "Google Announced 'Lens' for Advanced Image Recognition in Smartphones." *Fast Company*, May 17, 2017, https://news.fastcompany.com/google-announced-lens-tool-for-advanced-image-recognition-in-smartphones-4038050.

Surden, Harry, and Mary-Anne Williams. "Technological Opacity, Predictability, and Self-Driving Cars." *Cardozo Law Review* 38 (2016): 121–181.

Swisher, Kara, and Chris Urmson. "Full Transcript: Self-driving Car Engineer Chris Urmson on Recode Decode." *Recode*, September 8, 2017, www.recode.net/2017/9/8/16278566/transcript-self-driving-car-engineer-chris-urmson-recode-decode.

Tagg, John. "Mindless Photography." In *Photography: Theoretical Snapshots*. Edited by J. J. Long, Andrea Noble, and Edward Welch, 16–30. London: Routledge, 2009.

Takahashi, Dean. "6D.ai Creates Platform for a 3D Map of the World." *Venture Beat*, August 27, 2019, https://venturebeat.com/2019/08/27/6d-ai-creates-platform-for-a-3d-map-of-the-world/.

Tango. "Motion Tracking." https://developers.google.com/tango/overview/motion-tracking.

Tango. "Overview: Events." https://developers.google.com/tango/overview/events.

Tao, Li. "Facial Recognition Snares China's Air Con Queen Dong Mingzhu for Jay-walking, But It's Not What It Seems." *South China Morning Post*, November 23, 2018, www.scmp.com/tech/innovation/article/2174564/facial-recognition-catches-chinas-air-con-queen-dong-mingzhu.

Terranova, Tiziana. "Attention, Economy and the Brain." *Culture Machine* 13 (2012), https://culturemachine.net/wp-content/uploads/2019/01/465-973-1-PB.pdf.

Thrift, Nigel. "Remembering the Technological Unconscious by Foregrounding Knowledges of Position." *Environment and Planning D: Society and Space* 22 (2004): 175–190.

Tiidenberg, Katrin, and Edgar Gómez Cruz. "Selfies, Image and the Re-making of the Body." *Body & Society* 21, no. 4 (2015): 77–102.

Torralba, Antonio, and Alexei A. Efros. "Unbiased Look at Dataset Bias." In *Proceedings of the Conference on Computer Vision and Pattern Recognition (CVPR)*, 1521–1528. New York: IEEE, 2011.

Torres, Juan Carlos. "Two New Systems Can Help Driverless Cars to See Better." *Slash Gear*, December 23, 2015, www.slashgear.com/two-new-system-can-help-driverless-cars-see-better-23419658/.

Uhlemann, Elisabeth. "Introducing Connected Vehicles [Connected Vehicles]." *IEEE Vehicular Technology Magazine* 10, no. 1 (2015): 23–31.

"Unmanned Aircraft Systems (UAS)." *Federal Aviation Administration*, www.faa.gov/uas/.

Urmson, Chris, Chris Baker, John Dolan, Paul Rybski, Bryan Salesky, William Whittaker, Dave Ferguson, and Michael Darms. "Autonomous Driving in Traffic: Boss and the Urban Challenge." *AI Magazine*, Summer (2009): 17–28.

Urry, John. *Mobilities*. Cambridge: Polity, 2007.

van Dijck, José. *Mediated Memories in the Digital Age*. Stanford, CA: Stanford University Press, 2007.

van Dijck, José, Thomas Poell, and Martijn de Waal. *The Platform Society: Public Values in a Connective World*. London: Oxford University Press, 2018.

Van House, Nancy A. "Collocated Photo Sharing, Story-telling, and the Performance of Self." *International Journal of Human-Computer Studies* 67 (2009): 1073–1086.

Verhoeff, Nanna. "A Logic of Layers: Indexicality of iPhone Navigation in Augmented Reality." In *Studying Mobile Media: Cultural Technologies, Mobile Communication, and the iPhone*. Edited by Larissa Hjorth, Jean Burgess, and Ingrid Richardson, 118–132. New York: Routledge, 2012.

Verhoeff, Nanna. *Mobile Screens: The Visual Regime of Navigation*. Amsterdam: Amsterdam University Press, 2012.

Villi, Mikko. "Visual Chitchat: The Use of Camera Phones in Visual Interpersonal Communication." *Interactions: Studies in Communication and Culture* 3, no. 1 (2012): 39–54.

Villi, Mikko. "Visual Mobile Communication on the Internet: Patterns in Publishing and Messaging Camera Phone Photographs." In *Mobile Media Practices, Presence and Politics: The Challenge of Being Seamlessly Mobile*. Edited by Kathleen M. Cumiskey and Larissa Hjorth, 214–227. New York: Routledge, 2013.

Virilio, Paul. *The Vision Machine*. Translated by Julie Rose. London and Bloomington, IN: BFI/Indiana University Press, 1994.

Walker Rettberg, Jill. *Seeing Ourselves Through Technology: How We Use Selfies, Blogs and Wearable Devices to See and Shape Ourselves*. Houndmills, Basingstoke: Palgrave Macmillan, 2014.

Wang, Xiaosong, Yifan Peng, Le Lu, Zhiyong Lu, Mohammadhadi Bagheri and Ronald M. Summers. "ChestX-ray8: Hospital-scale Chest X-ray Database and Benchmarks on Weakly-Supervised Classification and Localization of Common Thorax Diseases." *Computer Vision Foundation* (2017), 2097–2016, http://openaccess.thecvf.com/content_cvpr_2017/papers/Wang_ChestX-ray8_Hospital-Scale_Chest_CVPR_2017_paper.pdf.

Wang, Zhou. "Is China's Grassroots Social Order Program Running Out of Money?" *Sixth Tone*. Translated by Kilian O'Donnell, June 1, 2018, www.sixthtone.com/news/1002393/is-chinas-grassroots-social-order-project-running-out-of-money%3F.

Wesch, Michael. "YouTube and You: Experiences of Self-Awareness in the Context Collapse of the Recording Webcam." *Explorations in Media Ecology* 8, no. 2 (2009): 19–34.

Whitwam, Ryan. "How Google's Self-Driving Cars Detect and Avoid Obstacles." *ExtremeTech*, September 8, 2014, www.extremetech.com/extreme/189486-how-googles-self-driving-cars-detect-and-avoid-obstacles.

"Why Should You Fly Freestyle at 800mW? | FPV." YouTube video, 0:15. "Mr Steele," July 28, 2018, www.youtube.com/watch?v=bBb_kSO3vTo&t=105s.

Whyte, William H. *The Social Life of Small Urban Spaces*. Washington, DC: The Conservation Foundation, 1980.

Wiessner, Siegfried. "The Public Order of the Geostationary Orbit: Blueprints for the Future." *Yale Journal of World Public Order* 9 (1983): 217–274.

Wiggers, Kyle. "Microsoft Releases Windows Vision Skills Preview to Streamline Computer Vision Development." *VentureBeat*, April 30, 2019, https://venturebeat.com/2019/04/30/microsoft-releases-windows-vision-skills-preview-to-streamline-computer-vision-development/.

Wilcox, Christie. "Why FaceApp's Selfie Filters Work So Well, and Why They Don't." *Gizmodo*, May 10, 2017, www.gizmodo.com.au/2017/05/why-faceapps-selfie-filters-work-so-well-and-why-they-dont/.

Wilken, Rowan, and Gerard Goggin, eds. *Mobile Technology and Place*. New York: Routledge, 2012.

Wilken, Rowan, and Julian Thomas. "Maps and the Autonomous Vehicle as a Communication Platform." *International Journal of Communication* 13 (2019), https://ijoc.org/index.php/ijoc/article/view/8450.

Williams, Alex. "The Power of Data Exhaust." *TechCrunch*, May 26, 2013, http://techcrunch.com/2013/05/26/the-power-of-data-exhaust/.

Wilson, Matthew W. "Critical GIS." In *Key Methods in Geography*, third edition. Edited by Nicholas Clifford, Shaun French, and Gill Valentine, 285–301. London: Sage, 2016.

Wolff, Josephine. "China's Push Towards Facial Recognition Technology." *China – US Focus*, March 12, 2019, www.chinausfocus.com/finance-economy/chinas-push-towards-facial-recognition-technology.

Wynne, Brian. "Risk and Social Learning: Reification to Engagement." In *Social Theories of Risk*. Edited by S. Krimsky and D. Golding, 275–300. Westport, CT: Praeger, 1992.

Zagoruyko, Sergey, Adam Lerer, Tsung-Yi Lin, Pedro O. Pinheiro, Sam Gross, Soumith Chintala and Piotr Dollár. "A Multipath Network for Object Detection." *arXiv preprint* (2016), https://arxiv.org/pdf/1604.02135.pdf.

Zhu, Xiangxin, and Deva Ramanan. "Face Detection, Pose Estimation, and Landmark Localization in the Wild." *International Conference on Computer Vision and Pattern Recognition (CVPR)*, 2879–2886. New York: IEEE, 2012.

Zylinska, Joanna. *Nonhuman Photography*. Cambridge, MA: MIT Press, 2017.

INDEX

Printed and bound by CPI Group (UK) Ltd, Croydon, CR0 4YY

23/10/2024

01778228-0001